Springer Tracts in Modern Physics
Volume 204

Managing Editor: G. Höhler, Karlsruhe

Editors: J. Kühn, Karlsruhe
Th. Müller, Karlsruhe
A. Ruckenstein, New Jersey
F. Steiner, Ulm
J. Trümper, Garching
P. Wölfle, Karlsruhe

Available online at
SpringerLink.com

Starting with Volume 165, Springer Tracts in Modern Physics is part of the [SpringerLink] service. For all customers with standing orders for Springer Tracts in Modern Physics we offer the full text in electronic form via [SpringerLink] free of charge. Please contact your librarian who can receive a password for free access to the full articles by registration at:

springerlink.com

If you do not have a standing order you can nevertheless browse online through the table of contents of the volumes and the abstracts of each article and perform a full text search.

There you will also find more information about the series.

Springer Tracts in Modern Physics

Springer Tracts in Modern Physics provides comprehensive and critical reviews of topics of current interest in physics. The following fields are emphasized: elementary particle physics, solid-state physics, complex systems, and fundamental astrophysics.

Suitable reviews of other fields can also be accepted. The editors encourage prospective authors to correspond with them in advance of submitting an article. For reviews of topics belonging to the above mentioned fields, they should address the responsible editor, otherwise the managing editor. See also springeronline.com

Peter Sigmund

Stopping of Heavy Ions

A Theoretical Approach

With 43 Figures

 Springer

Prof. Peter Sigmund
Physics Department
University of Southern Denmark
5230 Odense M
Denmark

Library of Congress Control Number: 200410790

Physics and Astronomy Classification Scheme (PACS): 41.5.-i; 61.80.Jh; 81.40.Wx

ISSN print edition: 0081-3869
ISSN electronic edition: 1615-0430
ISBN 3-540-22273-1 Springer Berlin Heidelberg New York

Springer is a part of Springer Science+Business Media

springeronline.com

Typesetting: Author using a Springer LaTeX macro package
Cover concept: eStudio Calamar Steinen
Cover production: *design & production* GmbH, Heidelberg

Printed on acid-free paper   SPIN: 11015697   57/3141/YL   5 4 3 2 1 0

Preface

Accelerators for charged particles with a very wide range of masses and charge states have an increasing application range in fundamental physics research, in medical radiology, in materials science and engineering, in micro and nano science and technology, in nuclear fission and fusion technology, and in mass spectrometry. The range of kinetic energies reached at existing accelerators extends from a few electron volts per particle into the 10^{12} electron volt regime. Effective and successful application of such particle beams requires detailed quantitative knowledge of the penetration of charged particles through matter, a field commonly categorized under the discipline of atomic physics.

Following the quest for critical data tabulations, the International Commission on Radiation Units and Measurements (ICRU) has issued tables on stopping powers and ranges for electrons in 1984 (ICRU Report 37) and for protons and alpha particles in 1993 (ICRU Report 49). A followup for ions heavier than helium was commissioned subsequently, and a report committee chaired by the author of this monograph has delivered a draft report which has passed the reviewing procedure of the commission and is going to be published by Oxford University Press in a forthcoming issue of the Journal of the ICRU. That report presents a survey of stopping theory for heavy ions and of experimental techniques and results, a critical discussion of available tabulations and computer codes, and stopping tables for numerous elements and compounds.

Experimental and theoretical activity in heavy-ion stopping has increased during the past decade, stimulated in part by the ICRU report committee. This, in turn, accentuated the need for an extensive review of the field, and especially a summary of the basic theoretical concepts. While a monograph of the breadth of Niels Bohr's legendary review from 1948 has not appeared ever since, the theory part in the forthcoming ICRU report should be able to serve as a useful guide on available knowledge and future research in the field of stopping of heavy ions, if it were to appear here and now. However, production of the full ICRU report has to follow a schedule agreed to between commission and publisher, and this is still going to take some time. For this reason it was found desirable to publish part of the report now. Duplicate publication appears justified since a monograph in Springer Tracts of Modern Physics is likely to attract a rather different readership than an ICRU report.

The structure of this monograph bears evidence of its origin. Considerable care has been devoted to precise definitions of all pertinent physical quantities, and an attempt has been made to present a balanced view of important developments. Although not a textbook, the monograph should also become useful in a university course on particle penetration. Detailed derivations are not given, but with the exception of elementary common concepts, all quoted results are supported by references to the literature[1]. I tried to select references according to the guiding principle to identify the first or the best research paper or the latest review. I am afraid that this did not prevent references to my own work from being overrepresented. My apologies in advance.

Also the topical coverage is influenced by the primary task issued by the ICRU, with a heavy emphasis on mean energy loss and fluctuation of swift, not too heavy ions in disordered media. Multiple scattering has been included mainly as a phenomenon affecting stopping measurements, while related phenomena such as stopping of channeled ions or of ions interacting with surfaces are mentioned rather briefly.

The author is grateful to the managements of three organizations involved, the ICRU, Springer, and Oxford University Press, for efficiently removing all obstacles toward publication of this material both as a monograph in Springer Tracts and as two chapters in the forthcoming ICRU report. Differences between the two versions are minute, and mainly motivated by the need to make this stand-alone version self-contained. Where appropriate, reference has been made to material contained in the full report.

The author's thanks go to Mitio Inokuti who initiated this project as an ICRU sponsor and took an active interest in its completion, and to several commission members, in particular Paul DeLuca and Stephen Seltzer for a great reviewing effort. Thanks are due to René Bimbot, Hans Geissel and Helmut Paul for extensive feedback to the text, and to Andreas Schinner, with whom I share responsibility for all aspects of binary stopping theory which forms the scientific basis of the tabulations in the forthcoming ICRU report. Valuable advice has been received from Nestor Arista, Martin Berger, Lev Glazov and Allan H. Sørensen, all of whom put considerable effort into reading several drafts. I am particularly grateful to two scientists who are no longer with us: Niels Bohr, whose insight and vision have had an all-dominating influence on the entire field of particle penetration through matter, and to Jens Lindhard, whose guidance, both directly and by example, has been instrumental for my activity in this area.

This work has been supported by the Danish Natural Science Research Council (SNF).

Odense,   *Peter Sigmund*
May 2004

---

[1]The bibliography has been closed on 1 October 2003.

Contents

List of Symbols and Abbreviations

a	adiabatic radius v/ω.
a	screening radius of interatomic potential.
a_0	Bohr radius.
A_1	mass number of projectile ion.
A_2	mass number of target atom.
a_{sc}	atomic screening radius.
B_j	Bethe parameter $2mv^2/\hbar\omega_j$.
c	speed of light.
C	constant 1.1229 in Bohr logarithm.
CKT	convergent kinetic theory.
csda	continuous-slowing-down approximation.
$-\mathrm{d}E/\rho\mathrm{d}\ell$	mass stopping force.
$-\mathrm{d}E/\mathrm{d}\ell$	stopping force, stopping power.
$-\mathrm{d}E/\mathrm{d}x$	stopping force, stopping power.
ΔE_{peak}	most probable energy loss.
$\mathrm{d}\mathcal{P}$	probability.
$\mathrm{d}\sigma$	differential cross section.
E	kinetic energy of projectile.
E_0	initial energy of projectile.
E/A_1	specific energy.
$F(\alpha,\ell)$	multiple-scattering profile.
$F(\Delta E,\ell)$	energy-loss profile.
$f(\eta)$	function determining differential cross section in Lindhard units.
f_j	oscillator strength of frequency ω_j.
$\hbar$	Planck's constant divided by 2π.
I	mean excitation energy.
I_1	I-value entering straggling.
Im	imaginary part.
J_n	Bessel function of the first kind.
k	absorption coefficient.
ℓ	path length.
L	stopping number.
L_0	stopping number omitting shell correction.

L_j	stopping number for jth shell.
LSS	Theory of ion ranges by Lindhard, Scharff and Schiøtt.
m	rest mass of electron.
m	exponent entering power cross section.
M_0	reduced mass.
M_1	rest mass of projectile ion.
M_2	rest mass of target atom.
n	number of target atoms per volume.
n	refractive index.
n_e	electron density.
$f(\omega), f'(\omega)$	oscillator-strength spectrum.
P	projectile momentum.
p	exponent in low-velocity stopping force.
p	impact parameter.
P_ℓ	Legendre polynomial.
$P(v, q_1)$	charge fraction.
q	exponent entering nuclear scattering cross section.
q_1	ion charge number.
$q_{1,\mathrm{eff}}$	effective-charge number.
$\mathbf{Q}, Q_{IJ}$	matrix governing charge fractions.
R	range, range along the path.
$\boldsymbol{R}$	vector range.
$\boldsymbol{r}$	position vector.
R	internuclear distance.
R	Rydberg energy.
Re	real part.
$R_\perp$	lateral range.
$r_{\min}$	distance of closest approach.
R_p	projected range.
R_p/R	projected-range correction, detour factor.
r_s	Wigner-Seitz radius.
S	stopping cross section.
S_e	electronic stopping cross section.
$s(\varepsilon)$	stopping cross section in Lindhard dimensionless units.
$\mathbf{S}, S_{IJ}$	stopping matrix.
S_n	nuclear stopping cross section.
t	time.
U	ionization energy.
u	screening function.
u	atomic mass unit.
UCA	unitary convolution approximation.
v	projectile speed.
$\boldsymbol{v}$	projectile velocity.
v_0	Bohr velocity.

$\boldsymbol{v}_e$	velocity of target electron.
W	straggling parameter.
w	energy transfer in single event.
w_1	limiting energy loss between multiple- and single-collision regime.
w_J	energy transfer to Jth loss channel.
w_{max}	maximum energy loss in single event.
$W_n(E)$	Nuclear-straggling parameter.
x	penetration depth.
Z_1	atomic number of projectile ion.
Z_2	atomic number of target atom.

α	total deflection angle.
α	fine structure constant $1/137$.
$\langle\ldots\rangle$	beam average.
β	v/c.
ΔE	energy loss at given pathlength.
$\langle\Delta E\rangle$	mean energy loss.
$\Delta E_{\pm 1/2}$	right and left halfwidth of energy-loss spectrum.
δ_ℓ	phase shift.
$\Delta L_{\mathrm{Bloch}}$	Bloch correction to stopping number.
$\Delta L_{\mathrm{invBloch}}$	inverse-Bloch correction to stopping number.
$\Delta\boldsymbol{u}$	velocity change in single collision.
$\Delta\boldsymbol{v}$	velocity change after penetration.
ϵ_0	vacuum permittivity.
$\varepsilon(\boldsymbol{k},\omega)$	Lindhard function.
ϵ	Lindhard dimensionless energy unit.
η	scaled nuclear energy loss.
γ	Euler's constant $= 0.5772$.
γ	$1/\sqrt{1-v^2/c^2}$.
γ^2	effective-charge fraction.
γ	energy-transfer factor $4M_1 M_2/(M_1 + M_2)^2$.
κ	Bohr's kappa parameter $2Z_1 v_0/v$.
λ	coefficient entering nuclear scattering cross section.
ν	species index.
ϕ	scattering angle in single event.
ψ	digamma function, $\psi(x) = \mathrm{d}\ln\Gamma(x)/\mathrm{d}x$.
ρ	mass density of medium.
σ_J	cross section for energy loss w_J.
$\sigma^{(1)}$	transport cross section determining stopping force.
$\sigma_B(k,\kappa)$	transport cross section determining energy loss and angular deflection.
$\boldsymbol{\sigma}, \sigma_{IJ}$	cross-section matrix.

$\sigma(s)$	transport cross section determining energy-loss spectrum.
ω	resonance frequency.
Ω^2	straggling, energy-loss straggling, variance of energy-loss profile.
Ω_B^2	Bohr straggling.
ω_j	resonance frequency of jth shell.
Ω_R^2	range straggling, variance of range profile.
ξ	Bohr parameter $mv^3/Z_1 v_0 \hbar\omega$.

1 Introduction

Quantitative information on the penetration of swift ions through matter, in particular the systematics of energy loss, is of considerable interest in basic science, in medicine and in technology. Until the middle of the past century, studies of charged-particle penetration were stimulated almost exclusively by the needs of fundamental physics research, but applications in other areas gradually became important. Table 1.1 gives an impression of the present range of applicability of accelerators for light and heavy ions covering a very wide energy range from the eV to the TeV regime.

Early studies of charged-particle penetration were stimulated by experiments on gas discharges toward the end of the 19'th century, but experimental possibilities were greatly enhanced after the discovery of radioactivity, in particular the pioneering work by E. Rutherford and coworkers in the beginning of the 20'th century. Pioneering theoretical studies by J. J. Thomson and N. Bohr date back to the same time. Subsequently, after the development of quantum mechanics, quantum theory of particle stopping was developed by H. Bethe, C. Møller, F. Bloch and others.

Table 1.1. Application areas for stopping data.

Physics and chemistry:	Cosmic radiation, radioactivity, accelerators, detectors etc.
Materials science and technology:	Analysis, modification, radiation damage
Fusion technology:	High energy density
Microelectronic devices :	Development, fabrication, control
Radiation medicine:	Diagnostics, therapy, radiation damage
Bio-, geo-, environmental sciences:	Mass spectrometry

Until the mid 1960s, experimental activities in the area focused on the penetration of light charged particles such as electrons and positrons, protons and numerous other low-charge particles. This was motivated by the needs of nuclear and particle physics. Moreover, options for experimental research on the penetration of particles heavier than helium were very limited in terms of available species and beam energies. This situation changed rapidly with new generations of ion sources and accelerators becoming available.

Until the early 1990s, theoretical research on particle stopping likewise focused on light penetrating particles, with the exception of pioneering work by Bohr (1940, 1941) on slowing-down of fission fragments and seminal work by Bohr and Lindhard (1954) on charge states of swift heavy ions. To this adds a large amount of studies on low-energy ion implantation.

The relatively weak Coulomb interaction of light particles with the electrons of the stopping medium allowed the application of well-developed concepts from quantum mechanical perturbation theory. Conversely, the passage of a heavy ion through a solid or gaseous material represents a strong intrusion that cannot generally be expected to be described adequately as a weak perturbation of the medium. Moreover, heavy ions are composite particles carrying electrons except at high speed, and their interaction with bound and free electrons in the stopping medium is a problem of considerable complexity involving a number of processes that are absent or less significant in the case of light projectiles.

Quantitative understanding of heavy-ion stopping was hampered for a very long time by two fundamental problems, the so-called charge-state paradox and the Barkas-Andersen effect. The charge-state paradox dates back to measurements by Lassen (1951a,b) where it was found that fission fragments travelling through solids carried significantly higher charges than when travelling through gaseous media, while the accompanying energy loss was almost the same for the two media. This was unexpected since the stopping force on a swift ion was thought to be proportional to the square of its charge, as was extrapolated from Bethe's theory for the stopping force on a *point charge.*

The Barkas-Andersen effect dates back to measurements of Smith *&* al. (1953) and subsequent work by the same group which demonstrated deviations from a strict proportionality of the stopping force to the square of the charge. This phenomenon was ascribed to higher-order perturbations and quantified by a contribution proportional to the third power of the charge of the penetrating particle. This contribution, verified for protons, alpha particles and lithium ions (Andersen *&* al., 1977), was expected to become increasingly pronounced for heavier ions. Reaching consensus on the theoretical understanding of the Barkas-Andersen effect took several decades for light projectiles. Extrapolations toward heavier projectiles were mostly speculative during this period.

As a result, stopping theory for heavy ions was quite fragmentary until around 1995. Theoretical predictions were available for comparatively slow ions (Firsov, 1959; Lindhard and Scharff, 1961) and for not too heavy ions at speeds approaching the velocity of light. For practical purposes, stopping parameters had to be extracted from empirical inter- and extrapolations (Steward and Wallace, 1966; Northcliffe and Schilling, 1970; Ziegler, 1980; Hubert *&* al., 1980; Ziegler *&* al., 1985; Hubert *&* al., 1990).

This situation changed rapidly after 1995. The theory by Bohr (1913), which had been considered until then to be mainly of historic interest, was rediscovered as a valuable and operative tool in the description of heavy-ion stopping. A new derivation by Lindhard and Sørensen (1996) reestablished the stopping formula of Bloch (1933) and demonstrated its superiority over the famous stopping formula of Bethe (1930) for the case of heavy ions. The same authors established a major improvement of the relativistic theory for stopping of heavy ions. Several independent schemes were developed to nonperturbatively describe the stopping of ions carrying electrons (Sigmund and Schinner, 2000, 2002; Grande and Schiwietz, 2002; Maynard & al., 2001, 2002; Arista, 2002).

With this development, it became possible to predict stopping forces without the use of fitting parameters. Moreover, several of the new schemes provided reasonable explanations of the charge-state paradox as well as predictions of the Barkas-Andersen effect and its dependence on the atomic number of the projectile.

The present monograph is intended to describe these developments and our present knowledge of the stopping of ions from lithium upward.

Prime parameters characterizing the penetration of charged particles are the mean energy loss per unit path length, i.e., the stopping power or stopping force, and its fluctuation, called energy-loss straggling. These quantities determine the penetration depth (range) and its fluctuation (range straggling) as well as the energy-deposition profile. Since particles are more or less deflected from their initial direction of motion, also the spatial and directional distribution, the multiple-scattering profile, is of interest.

This monograph focuses on mean energy loss and straggling. Multiple scattering as a separate phenomenon is treated in much less detail, but attention is given to the effect of multiple scattering on stopping measurements. Ranges are treated rather superficially.

As mentioned in the preface, this is not a textbook, and this implies that very few explicit derivations are given even of the most central findings. While a textbook on particle penetration by the present author has been in preparation for many years, the best available reference is still the monograph of Bohr (1948). Other summaries and introductory texts such as Fano (1963); Sigmund (1975); Inokuti (1971) and Bonderup (1981) more or less focus on light particles but may nevertheless help newcomers to find their way into stopping theory for heavy ions. Selected aspects have been discussed in reviews and monographs by Ahlen (1980), Kumakhov and Komarov (1981) and Ziegler & al. (1985). Basic experimental aspects have been discussed in classic reviews by Northcliffe (1963) and Betz (1972).

References

Ahlen, S. P. (1980). "Theoretical and experimental aspects of the energy loss of relativistic heavily ionizing particles," Rev. Mod. Phys. **52**, 121–173.

Andersen, H. H., Bak, J. F., Knudsen, H. and Nielsen, B. R. (1977). "Stopping power of Al, Cu, Ag and Au for MeV hydrogen, helium, and lithium ions. Z3 1 and Z4 1 proportional deviations from the Bethe formula," Phys. Rev. A **16**, 1929.

Arista, N. R. (2002). "Energy loss of heavy ions in solids: non-linear calculations for slow and swift ions," Nucl. Instrum. Methods B **195**, 91–105.

Bethe, H. (1930). "Zur Theorie des Durchgangs schneller Korpuskularstrahlen durch Materie," Ann. Physik **5**, 324–400.

Betz, H. D. (1972). "Charge states and charge-changing cross sections of fast heavy ions penetrating through gaseous and solid media," Rev. Mod. Phys. **44**, 465–539.

Bloch, F. (1933). "Zur Bremsung rasch bewegter Teilchen beim Durchgang durch Materie," Ann. Physik **16**, 285–320.

Bohr, N. (1913). "On the theory of the decrease of velocity of moving electrified particles on passing through matter," Philos. Mag. **25**, 10–31.

Bohr, N. (1940). "Scattering and stopping of fission fragments," Phys. Rev. **58**, 654–655.

Bohr, N. (1941). "Velocity-range relation for fission fragments," Phys. Rev. **59**, 270–275.

Bohr, N. (1948). "The penetration of atomic particles through matter," Mat. Fys. Medd. Dan. Vid. Selsk. **18 no. 8**, 1–144.

Bohr, N. and Lindhard, J. (1954). "Electron capture and loss by heavy ions penetrating through matter," Mat. Fys. Medd. Dan. Vid. Selsk. **28 no. 7**, 1–31.

Bonderup, E. (1981). *Interaction of charged particles with matter* (Institute of Physics, Aarhus).

Fano, U. (1963). "Penetration of protons, alpha particles, and mesons," Ann. Rev. Nucl. Sci. **13**, 1–66.

Firsov, O. B. (1959). "A qualitative interpretation of the mean electron excitation energy in atomic collsions," Zh. Eksp. Teor. Fiz. **36**, 1517–1523, [English translation: Sov. Phys. JETP **9**, 1076-1080 (1959)].

Grande, P. L. and Schiwietz, G. (2002). "The unitary convolution approximation for heavy ions," Nucl. Instrum. Methods B **195**, 55–63.

Hubert, F., Bimbot, R. and Gauvin, H. (1990). "Range and stopping-power tables for 2.5 - 500 MeV nucleon heavy ions in solids," At. Data and Nucl. Data Tab. **46**, 1–213.

Hubert, F., Fleury, A., Bimbot, R. and Gardès, D. (1980). "Range and stopping power tables for 2.5 - 100 MeV/nucleon heavy ions in solids," Ann. de Phys. **5** S, 3–213.

Inokuti, M. (1971). "Inelastic collisions of fast charged particles with atoms and molecules- the Bethe theory revisited," Rev. Mod. Phys. **43**, 297–347.

Kumakhov, M. A. and Komarov, F. F. (1981). *Energy loss and ion ranges in solids* (Gordon and Breach, New York).

Lassen, N. O. (1951a). "Total charges of fission fragments as functions of the pressure in the stopping gas," Mat. Fys. Medd. Dan. Vid. Selsk. **26 no. 12**, 1–19.

Lassen, N. O. (1951b). "The total charges of fission fragments in gaseous and solid stopping media," Mat. Fys. Medd. Dan. Vid. Selsk. **26 no. 5**, 1–28.

Lindhard, J. and Scharff, M. (1961). "Energy dissipation by ions in the keV region," Phys. Rev. **124**, 128–130.

Lindhard, J. and Sørensen, A. H. (1996). "On the relativistic theory of stopping of heavy ions," Phys. Rev. A **53**, 2443–2456.

Maynard, G., Sarrazin, M., Katsonis, K. and Dimitriou, K. (2002). "Quantum and classical stopping cross-sections of swift heavy ions derived from the evolution with time of the Wigner function," Nucl. Instrum. Methods B **193**, 20–25.

Maynard, G., Zwicknagel, G., Deutsch, C. and Katsonis, K. (2001). "Diffusion-transport cross section and stopping power of swift heavy ions - art. no. 052903," Phys. Rev. A **63**, 052903-1–14.

Northcliffe, L. C. (1963). "Passage of heavy ions through matter," Ann. Rev. Nucl. Sci. **13**, 67–102.

Northcliffe, L. C. and Schilling, R. F. (1970). "Range and stopping power tables for heavy ions," Nucl. Data Tab. **A 7**, 233–463.

Sigmund, P. (1975). "Energy loss of charged particles in solids," C. H. S. Dupuy, ed., *Radiation damage processes in materials*, NATO Advanced Study Institutes Series, 3–117 (Noordhoff, Leyden).

Sigmund, P. and Schinner, A. (2000). "Binary stopping theory for swift heavy ions," Europ. Phys. J. D **12**, 425–434.

Sigmund, P. and Schinner, A. (2002). "Binary theory of electronic stopping," Nucl. Instrum. Methods B **195**, 64–90.

Smith, F. M., Birnbaum, W. and Barkas, W. H. (1953). "Measurements of meson masses and related quantities," Phys. Rev. **91**, 765–766.

Steward, P. G. and Wallace, R. (1966). "Calculation of stopping power and range-energy values for any heavy ion in nongaseous media," Technical report UCRL-17314, Univ. California, Berkeley.

Ziegler, J. F. (1980). "The stopping and ranges of ions in matter," J. F. Ziegler, ed., *Handbook of stopping cross-sections for energetic ions in all elements*, volume 5 of *The Stopping and Ranges of Ions in Matter*, 1–432 (Pergamon, New York).

Ziegler, J. F., Biersack, J. P. and Littmark, U. (1985). "The stopping and range of ions in solids," J. F. Ziegler, ed., *The Stopping and Ranges of Ions in Matter*, volume 1 of *The Stopping and Ranges of Ions in Matter*, 1–319 (Pergamon, New York).

2 Definitions

2.1 Mean Energy Loss

The central quantity characterizing particle stopping is the *stopping force* or *stopping power*[1]. For a point particle it is defined as the average loss of kinetic energy E per path length ℓ, $-\mathrm{d}E/\mathrm{d}\ell$. The minus sign defines the stopping force as a positive quantity.

The stopping force is related to the average change in momentum per path length according to

$$\frac{\mathrm{d}E}{\mathrm{d}\ell} = v\frac{\mathrm{d}P}{\mathrm{d}\ell}, \tag{2.1}$$

where v is the projectile speed, $P = M_1\gamma v$ the momentum, M_1 the projectile mass,

$$\gamma = \frac{1}{\sqrt{1 - \beta^2}}, \tag{2.2}$$

$$\beta = \frac{v}{c} \tag{2.3}$$

and c the speed of light[2].

For composite projectiles such as molecules and clusters for which energy can be transferred into internal degrees of freedom, the stopping force is defined by generalizing (2.1):

$$\frac{\mathrm{d}E}{\mathrm{d}\ell} = \sum_i v_i \left(\frac{\mathrm{d}P}{\mathrm{d}\ell}\right)_i \tag{2.4}$$

where the sum extends over all constituents of the projectile.

Equation (2.4) in principle also applies to dressed atomic ions, i.e., ions carrying electrons for which the definition of the stopping force becomes

---

[1] The two terms will be used synonymously. While 'stopping power' is the official nomenclature, 'stopping force' is more precise (Sigmund, 2000).

[2] The symbol γ has at least four well-established functions in the context of heavy-ion penetration. In addition to (2.2), it characterizes the maximum energy transfer in an elastic binary collision, the effective-charge fraction of a dressed ion, and Euler's constant. Established notation will be maintained here, but proper specification will be added whenever ambiguities could arise.

ambiguous due to projectile excitation and charge exchange. However, currently accessible experimental and theoretical accuracy is below the level where terms of order m/M_1 (m = electron rest mass) would be significant. Therefore, the stopping force on a dressed ion may safely be related to the momentum change of the *projectile nucleus,*

$$\frac{\mathrm{d}E}{\mathrm{d}\ell} = \left(v\frac{\mathrm{d}P}{\mathrm{d}\ell} \right)_{\text{nucleus}}. \tag{2.5}$$

The definitions (2.4) and (2.5) allow the inclusion of contributions from all significant energy-loss channels.

The stopping force is related to the velocity change in time – a quantity measured by some techniques – through

$$\frac{\mathrm{d}E}{\mathrm{d}\ell} = \frac{\mathrm{d}P}{\mathrm{d}t} = M_1\gamma^3\frac{\mathrm{d}v}{\mathrm{d}t}. \tag{2.6}$$

It is common practice to tabulate *Mass stopping forces*

$$-\frac{1}{\rho}\frac{\mathrm{d}E}{\mathrm{d}\ell}, \tag{2.7}$$

where ρ is the mass density of the target.

A parameter of fundamental significance on the atomic or molecular scale is the *stopping cross section S*, defined by

$$S = \sum_J w_J\sigma_J, \tag{2.8}$$

where the sum extends over all energy-loss channels and where w_J and σ_J denote the energy loss and pertinent cross section per target atom or molecule for the Jth channel.

The stopping cross section is related to the stopping force through

$$S = -\frac{1}{n}\frac{\mathrm{d}E}{\mathrm{d}\ell}, \tag{2.9}$$

where n is the number of target atoms or molecules per volume. The relation to the mass stopping force reads

$$\frac{1}{M_2}S = -\frac{1}{\rho}\frac{\mathrm{d}E}{\mathrm{d}\ell}, \tag{2.10}$$

where M_2 is the mass of a target atom (or molecule) if the target consists of only one type of atom or molecule. For polyatomic and polymolecular targets, (2.10) remains valid if S and M_2 are replaced by averages in accordance with the respective atomic or molecular abundances.

2.2 Energy-Loss Fluctuation

The energy loss ΔE at a given pathlength ℓ is a stochastic variable which obeys a statistical distribution $F(\Delta E, \ell)$ depending on path length and ion-target combination.

The mean energy loss is connected to the stopping force by

$$\langle \Delta E \rangle = \int \mathrm{d}(\Delta E)\, \Delta E\, F(\Delta E, \ell) = -\frac{\mathrm{d}E}{\mathrm{d}\ell}\, \ell, \tag{2.11}$$

provided that ℓ is small enough so that the variation of $\mathrm{d}E/\mathrm{d}\ell$ across the pathlength segment can be neglected.

The notation $\langle \dots \rangle$ introduced in (2.11) denotes an average over a beam, i.e., a large number of trajectories. This includes all relevant parameters characterizing individual projectiles such as energy, charge and excitation state and position in space.

The fluctuation in energy loss ('energy-loss straggling') after a given path length ℓ is described primarily by the variance

$$\Omega^2 = \left\langle \left(\Delta E - \langle \Delta E \rangle \right)^2 \right\rangle . \tag{2.12}$$

Ω^2 is proportional to ℓ if individual energy-loss events are statistically independent. The *straggling parameter* W is then defined by

$$W = \frac{1}{n}\frac{\mathrm{d}\Omega^2}{\mathrm{d}\ell}, \tag{2.13}$$

and W may be expressed in atomic-scale parameters by

$$W = \sum_{J} w_J^2 \sigma_j . \tag{2.14}$$

Despite the similarity with (2.8), the range of validity of (2.14) is more restricted. It does not describe *charge-exchange straggling*, and since it is based on Poisson statistics, limitations occur especially in crystals and quite generally in dense media (Sigmund, 1978, 1991).

The question of the *shape* of an energy-loss spectrum deserves special attention. For very thin layers the spectrum must reflect the cross sections governing individual scattering events with a pronounced peak at small energy transfers and a tail at the high-energy end. Skewness decreases with increasing thickness, and eventually the spectrum reaches gaussian shape when (Bohr, 1948)

$$\Omega \gg w_{\mathrm{max}}, \tag{2.15}$$

where w_{max} is the maximum energy loss in an individual event. At thicknesses for which the total energy loss amounts to a sizable fraction of the initial energy, the spectra skew again.

As the rate of energy loss is governed by the speed, the range of thicknesses within which the gaussian approximation is valid widens for heavy ions because of increasing energy at constant speed. The opposite trend is observed when the projectile is an electron. There the gaussian limit of an energy-loss profile is hardly ever reached.

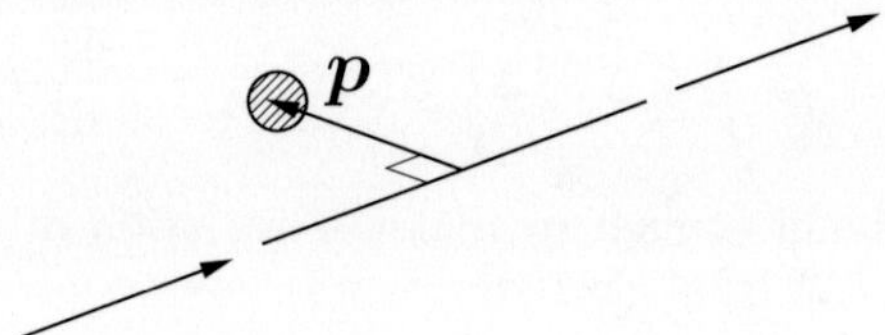

Fig. 2.1. Definition of impact parameter.

2.3 Impact-Parameter Dependence

The *impact parameter* in a collision denotes the distance between the incoming (straight-line) trajectory and the target when the latter is initially at rest (Fig. 2.1). While this is an inherently classical concept, it retains physical significance in the context of heavy-ion penetration. An impact parameter may refer to a target nucleus or a target electron. Unless stated otherwise, reference will be made here to the target nucleus. This is consistent with the so-called semiclassical picture where trajectories of nuclei participating in a collision are characterized by classical orbits, while electronic motion obeys the laws of quantum mechanics.

Beam averages defined in Sect. 2.2 can be interpreted as integrations over the *impact plane*, i.e., a plane perpendicular to the beam direction with the impact parameter being the radial coordinate.

The dependence on impact parameter of the energy loss must be considered whenever the stopping medium is inhomogeneous or anisotropic or when finite geometric dimensions cannot be ignored. Prominent examples are the stopping of a beam under grazing incidence on a flat surface, or of a beam incident on a single crystal under *channeling* conditions (cf. Sect. 11.2). Angular scattering of heavy ions - being governed by interactions with target nuclei - and the associated process of stopping by energy loss to recoil nuclei is another area for which knowledge of impact-parameter dependencies is vital.

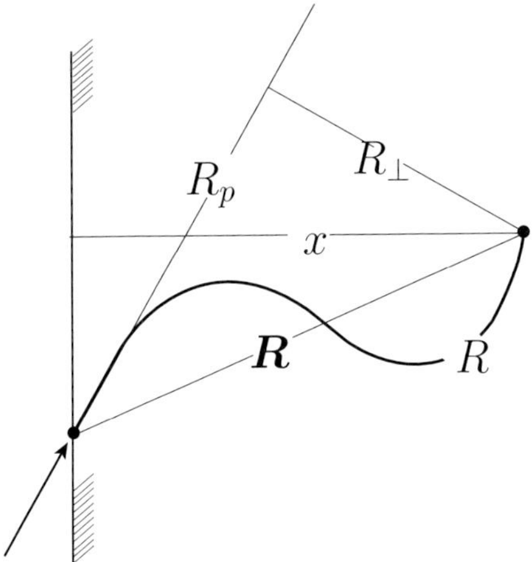

Fig. 2.2. Range concepts illustrated schematically.

2.4 Range

The *path length* ℓ specifies the length of a segment of the trajectory measured along the path. The pathlength between two points 1 and 2 is related to the stopping force by

$$\ell = \int_1^2 \mathrm{d}\ell = \int_{E_1}^{E_2} \frac{\mathrm{d}E}{\mathrm{d}E/\mathrm{d}\ell} = \int_{E_2}^{E_1} \frac{\mathrm{d}E}{nS(E)}, \tag{2.16}$$

provided that energy-loss fluctuations can be neglected. In the presence of small fluctuations, (2.16) is an approximate measure of the mean path length with the limits of integration specifying the end points of the path segment. In particular, the *range* or *range along the path* or csda (*continuous-slowing-down approximation*) range is given by

$$R = \int_0^{E_0} \frac{\mathrm{d}E}{nS(E)}, \tag{2.17}$$

where E_0 is the initial energy (Fig. 2.2). Equation (2.17) approximates the mean range along the path when straggling is small.

The variance in range or *range straggling* Ω_R^2 may be estimated from the formula of Bohr (1948)

$$\Omega_R^2 = \int_0^{E_0} \mathrm{d}E \, \frac{nW(E)}{[nS(E)]^3}, \tag{2.18}$$

again in the limit of low straggling.

In addition, the following range parameters (Lindhard *&* al., 1963b) are of interest, cf. Fig. 2.2,

- The *vector range* $\boldsymbol{R}$, specifying the vector distance from the starting point to the end point of a trajectory,
- The *projected range* R_p, representing the component of the vector range in the initial direction of motion,
- The *lateral range* $R_\perp$, representing the component of the vector range perpendicular to the initial direction of motion, and
- The *penetration depth* x, representing the component of the vector range along a given direction, e.g. the surface normal of a target with a plane surface.
- A useful dimensionless parameter is the *projected-range correction* or *detour factor* R_p/R which is always ≤ 1.

Range profiles as well as average ranges and variances can be associated with each of the range concepts defined above. These definitions override the approximate relationships (2.17) and (2.18) in the presence of significant straggling and/or angular scattering.

2.5 Radiation Effects

Radiation effects such as damage and ionization are characterized primarily by *energy deposition profiles*. Both one- and three-dimensional profiles are of interest, i.e., the energy deposited per depth or per volume by the ion and by all secondary particles such as recoiling target atoms and secondary electrons. It is necessary to distinguish between different modes of energy deposition (Lindhard *&* al., 1963a). Unlike range distributions which govern the statistical distribution of a single point, i.e., the end point of the ion trajectory, damage distributions characterize a collision or ionization cascade determined by the fate of a multitude of moving particles. Nevertheless, in the average over a large number of trajectories, damage and ionization profiles are governed by distribution functions analogous to and quantitatively not too different from the corresponding range distributions.

Figure 2.3 shows the simplest estimate of an energy deposition profile, found by ignoring straggling and angular scattering. Integration of (2.9) specifies the projectile energy $E = E(\ell)$ as a function of pathlength as the inverse of the relation

$$\ell = \ell(E) = \int_E^{E_0} \frac{\mathrm{d}E'}{nS(E')}. \tag{2.19}$$

The energy deposited in ionization per pathlength $F_{\mathrm{ioniz}}(\ell)$ may then be expressed as a function of pathlength by

$$F_{\mathrm{ioniz}}(\ell) = nS_{\mathrm{ioniz}}\big(E(\ell)\big), \tag{2.20}$$

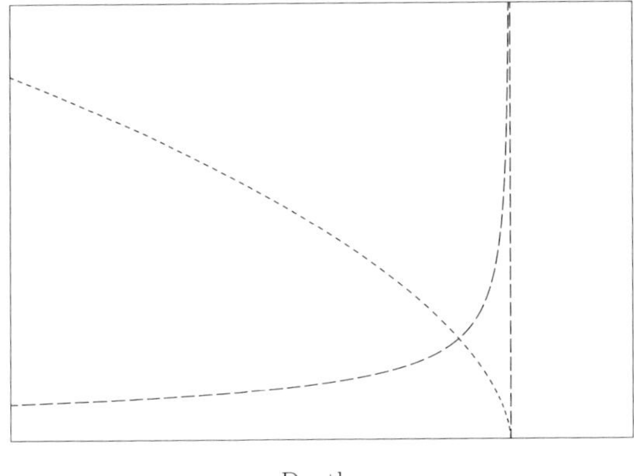

Depth

Fig. 2.3. Beam energy (short-dashed line) and deposited energy (long-dashed line) *versus* penetration depth in the absence of energy-loss straggling (schematic).

where S_{ioniz} symbolizes the contribution of ionization processes to the total stopping cross section (*Bragg curve*). Similar relations may be written down for other radiation effects.

2.6 Angular Scattering

Angular scattering affects stopping measurements mainly in two ways,

- Beam particles deflected away from the detecting device may give rise to distortion of the measured energy-loss spectrum and of all averages,
- Differences between travelled path length and penetration depth through a layer give rise to a detour factor in energy loss and, more seriously, in range.

While the second feature is a key ingredient in standard range theory, consideration of the first one plays a role in all analysis of stopping data. At this point, pertinent notation is introduced.

The probability for angular deflection into a solid angle $\mathrm{d}^2\phi = 2\pi \sin\phi \mathrm{d}\phi$ is given by

$$\mathrm{d}\mathcal{P} = n\ell K(\phi)\mathrm{d}^2\phi, \qquad (2.21)$$

if the travelled pathlength ℓ is small enough so that $\mathrm{d}\mathcal{P} \ll 1$, where $K(\phi)$ is the *differential scattering cross section* and ϕ the deflection angle in the laboratory frame of reference.

With $\mathrm{d}\mathcal{P}$ increasing, multiple angular deflections become increasingly important. The distribution in total scattering angle α is then described by a distribution $F(\alpha, \ell)\mathrm{d}^2\alpha$ which typically approaches the single-scattering profile (2.21) at large angles but takes on a gaussian-like shape around $\alpha = 0$.

Complications arise from the fact that angular deflection and energy loss are correlated (cf. sect. 15).

References

Bohr, N. (1948). "The penetration of atomic particles through matter," Mat. Fys. Medd. Dan. Vid. Selsk. **18 no. 8**, 1–144.

Lindhard, J., Nielsen, V., Scharff, M. and Thomsen, P. V. (1963a). "Integral equations governing radiation effects," Mat. Fys. Medd. Dan. Vid. Selsk. **33 no. 10**, 1.

Lindhard, J., Scharff, M. and Schiøtt, H. E. (1963b). "Range concepts and heavy ion ranges," Mat. Fys. Medd. Dan. Vid. Selsk. **33 no. 14**, 1.

Sigmund, P. (1978). "Statistics of particle penetration," Mat. Fys. Medd. Dan. Vid. Selsk. **40 no. 5**, 1–36.

Sigmund, P. (1991). "Statistics of charged-particle penetration," A. Gras-Marti, H. M. Urbassek, N. Arista and F. Flores, eds., *Interaction of charged particles with solids and surfaces*, volume 271 of *NATO ASI Series*, 73–144 (Plenum Press, New York).

Sigmund, P. (2000). "Stopping power: wrong terminology," ICRU News **1**, 17.

3 Units, Fundamental Constants and Conversion Factors

3.1 Units

While the use of gaussian units is most common in the theoretical literature on particle stopping, the SI system of units is clearly preferrable from a user's point of view. A third option, atomic units, is frequently encountered. This section serves to provide pertinent expressions for key parameters in these three systems of units and suitable conversion factors.

Table 3.1 specifies expressions for Bohr radius, Rydberg energy and Bohr velocity.

Table 3.1. Fundamental constants entering stopping parameters: Numerical values (IUPAP, 1987) and expressions in three systems of units.

Name	Symbol	Value	SI	Gaussian	a. u.
Bohr radius	a_0	0.052917 nm	$4\pi\epsilon_0\hbar^2/me^2$	$\hbar^2/me^2$	1
Rydberg energy	R	13.6057 eV	$e^2/8\pi\epsilon_0 a_0$	$e^2/2a_0$	0.5
Bohr velocity	v_0	$c/137.036$	$e^2/4\pi\epsilon_0\hbar$	$e^2/\hbar$	1

3.2 Examples

The use of different units is illustrated on the Coulomb factor in the standard expression for the electronic stopping cross section of a point charge which, in gaussian units, reads

$$S = \frac{4\pi Z_1^2 Z_2 e^4}{mv^2} L, \tag{3.1}$$

where L denotes the dimensionless *stopping number* (which would read $L = \ln(2mv^2/I)$ in case of the Bethe formula) and Z_1, Z_2 denote the atomic numbers of projectile and target, respectively.

The same relation in SI units reads

$$S = \frac{Z_1^2 Z_2 e^4}{4\pi\epsilon_0^2 m v^2} L \tag{3.2}$$

but is rarely if ever encountered in the literature. In terms of fundamental constants (IUPAP, 1987; NIST, 2001) listed in Table 3.1 the same relation reads

$$S = 4\pi Z_1^2 Z_2 \left(\frac{v_0}{v}\right)^2 L\, 2\mathrm{R}\, a_0^2. \tag{3.3}$$

In atomic units this reduces to

$$S = 4\pi Z_1^2 Z_2 \frac{L}{v^2}. \tag{3.4}$$

The atomic unit of S is equivalent to[1] 27.2 eV $\times$ $(0.0529$ nm$)^2$.

3.3 Conversion

In practical applications the kinetic energy per nucleon or per atomic mass unit E/A_1, or *specific energy* is more convenient than the velocity variable[2].

Mass stopping forces, (2.7), will be reported in units of MeV cm^2/mg and stopping cross sections in units of eV cm^2 per atom or molecule.

Pertinent relations for the most frequently occurring conversions are listed here. The specific energy is related to the projectile speed by

$$\frac{E}{A_1} = uc^2(\gamma - 1) = 931.49 \left(\frac{1}{\sqrt{1 - v^2/c^2}} - 1\right), \tag{3.5}$$

E/A_1 in MeV,

where u is the atomic mass unit 1.6605×10^{-27} kg. The inverse relation reads

$$\frac{1}{2} M_1 v^2 = E \frac{1 + E/2M_1 c^2}{(1 + E/M_1 c^2)^2}. \tag{3.6}$$

The mass stopping force relates to the stopping number as

$$-\frac{\mathrm{d}E}{\rho\mathrm{d}\ell} = 3.0705 \cdot 10^{-4} \frac{Z_1^2 Z_2}{A_2 \beta^2} L \tag{3.7}$$

---

[1] Caution is indicated with respect to the energy unit which is set to 2R here. This is consistent with the common practice of setting $\hbar = m = e = 1$. However, also the straight Rydberg energy R may be encountered as the energy unit, and this is not always noted explicitly in the pertinent literature.

[2] In general, no distinction will be made between energy per nucleon and energy per atomic mass unit because the numerical difference amounts to at most 0.25 % for stable isotopes of all elements from lithium upward.

$\mathrm{d}E/\rho\mathrm{d}\ell$ in MeV cm^2/mg,

and to the stopping cross section as

$$S = 1.6605\, A_2 \left(-\frac{\mathrm{d}E}{\rho\mathrm{d}\ell} \right) \tag{3.8}$$

S in 10^{-15}eV cm^2, $\mathrm{d}E/\rho\mathrm{d}\ell$ in MeV cm^2/mg.

The stopping cross section relates to the stopping number as

$$S = 5.0991 \cdot 10^{-4} \frac{Z_1^2 Z_2}{\beta^2} L \tag{3.9}$$

S in 10^{-15}eV cm^2,

and the stopping force in atomic units is related to the mass stopping force as

$$\frac{\mathrm{d}E}{\mathrm{d}\ell} = 0.36749\rho \left(\frac{\mathrm{d}E}{\rho\mathrm{d}\ell} \right) \tag{3.10}$$

ρ in g/cm^3,
$\mathrm{d}E/\rho\mathrm{d}\ell$ in MeV cm^2/mg.

Alternatively,

$$\frac{\mathrm{d}E}{\mathrm{d}\ell} = 100\,\rho\,\frac{\mathrm{d}E}{\rho\mathrm{d}\ell}. \tag{3.11}$$

$\mathrm{d}E/\mathrm{d}\ell$ in keV/µm, ρ in g/cm^3,
$\mathrm{d}E/\rho\mathrm{d}\ell$ in MeV cm^2/mg

in atomic and modified SI units, respectively.

Finally, the specific energy of an ion at the Bohr velocity v_0 is given by

$$\left(\frac{E}{A_1} \right)_{v=v_0} = 0.02480 \text{ MeV.} \tag{3.12}$$

3.4 Notation

In the stopping literature, the symbol S may be found to denote both stopping cross section, stopping force and mass stopping force, and the symbol $-\mathrm{d}E/\mathrm{d}x$ may be found to denote either stopping force or mass stopping force.

References

IUPAP (1987). *Symbols, units, nomenclature and fundamental constants in physics* (International Union of Pure and Applied Physics).
NIST (2001). "Fundamental physical constants," Technical report, National Institute of Standards and Technology, http://physics.nist.gov/cuu/Constants/.

4 General Considerations

4.1 Introductory Remarks

Early theoretical efforts on heavy-ion stopping date back to Bohr (1940) who pointed out the importance of screening due to projectile electrons in the slowing-down of fission fragments, and to Lamb (1940) and Knipp and Teller (1941) who studied the problem of charge equilibrium for penetrating heavy particles. The central role of the projectile charge as well as charge exchange in conjunction with stopping phenomena was discussed by Bohr and Lindhard (1954).

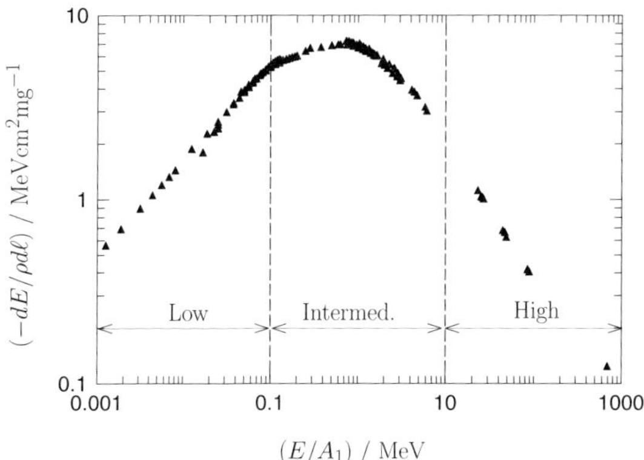

Fig. 4.1. Regimes of heavy-ion stopping illustrated by oxygen in aluminium. Data compiled by Paul (2003).

Since the appearance of the LSS theory (Lindhard & al., 1963) it has been common to divide heavy-ion stopping into three regimes (Fig. 4.1),

- a low-speed regime where the electronic stopping force is taken to be proportional to the projectile speed v and given roughly by the estimates of Lindhard and Scharff (1961) or Firsov (1959),
- a high-speed regime characterized by the Bethe (1930) formula, and

- an intermediate regime around and above the stopping maximum which has most often been characterized by a Bethe-type formula in conjunction with some effective ion charge (Northcliffe, 1963).

While this picture has been of some help in attempts to scale experimental data, the process of theoretical understanding and quantitative description in particular of the intermediate regime has been slow. Amongst a variety of reasons are the notorious problem of understanding the connection between ion charge and energy loss, in particular in connection with the so-called density effect, i.e., the distinct difference in measured equilibrium charge states between gaseous and solid media mentioned in Ch. 1, uncertainty about the role of the Barkas-Andersen effect (cf. Sect. 5.4), and lack of knowledge about the contribution of charge exchange and projectile excitation at intermediate velocities.

An important step forward was made by Brandt and Kitagawa (1982) who established an explicit connection between ion charge and stopping force. Several details of this theory are either too restrictive or obsolete, but its central feature, to let the ion charge enter via a partially screened Coulomb potential as suggested originally by Bohr (1948) in his famous review of particle penetration, has been common to all subsequent theoretical attacks on this problem.

In the quoted paper, Bohr pointed at the fact that the regimes of validity of classical-orbit models and of quantal perturbation theory are roughly complementary: The stopping of low-charge particles like electrons and protons at high speed, i.e., well above the Bohr velocity $v_0 = c/137$, is accurately described by the Bethe theory which treats projectile-target interaction by quantal perturbation theory to lowest order. Since the accuracy of this scheme deteriorates with increasing projectile charge and decreasing speed it seemed appropriate in an alternative approach to start at the opposite end, i.e. the classical limit, in an attempt to find a comprehensive theory of heavy-ion stopping. This led to the binary theory of stopping (Sigmund and Schinner, 2000) and various extensions.

Parallel developments in the theory of heavy-ion stopping include the so-called CKLT model by Maynard & al. (2001) and the convolution approximation by Grande and Schiwietz (1998, 2002), both geared toward intermediate to high velocities. A model by Lifschitz and Arista (1998) based on a generalization of the Friedel sum rule, initially designed for low-speed light-ion and antiproton stopping, has been extended to heavy ions (Arista, 2002). It is geared toward low and intermediate velocities and appears particularly promising for the low-speed range.

4.2 Classification

Energy-loss processes for charged particles may be classified roughly into five groups,

1. Excitation and ionization of target electrons,
2. Projectile excitation and ionization,
3. Electron capture,
4. Recoil loss ('nuclear stopping'),
5. Electromagnetic radiation.

For electrons only processes 1, 4, and 5 are of interest (ICRU, 1984). For light ions (ICRU, 1993) electromagnetic radiation (process 5) is negligible up to very high energies and process 1 dominates except at the low-speed end. This simple picture changes for heavier ions, where processes 2 and 3 cannot be neglected in general and, moreover, nuclear stopping becomes relatively more important at low and moderate velocities. Radiative processes become dominating at extremely high velocities.

Energy may also be transferred into

6. Nuclear reactions and
7. Chemical reactions.

It is debatable whether or not such processes should be categorized under stopping as projectiles may change identity. While chemical reactions do not affect the stopping process significantly at keV and higher energies, caution is indicated with regard to nuclear reactions. Although it makes little sense to relate the stopping force on the fragments of a disintegrated projectile to that on the mother nucleus, the energy-deposition profile in the stopping medium hinges on all these quantities. Thus, the preferred – and certainly most rigorous – strategy must be to incorporate nuclear reactions that change the identity of the projectile into an appropriate theory of radiation effects and to omit them from the stopping force.

4.3 Thomas-Fermi Estimates

The Thomas-Fermi model of the atom (Gombas, 1956) ignores the shell structure but provides useful estimates for the qualitative dependence of parameters characterizing an atom as a function of atomic number. Thomas-Fermi parameters read

- $Z^{-1/3} a_0$ for length,
- $Z^{2/3} v_0$ for speed,
- $Z^{4/3} R$ for energy per electron, and
- $Z R/\hbar$ for frequency,

for a neutral atom with atomic number Z. Estimates involving these parameters will be applied to quantities characterizing target (Z_2) and projectile (Z_1) in the following.

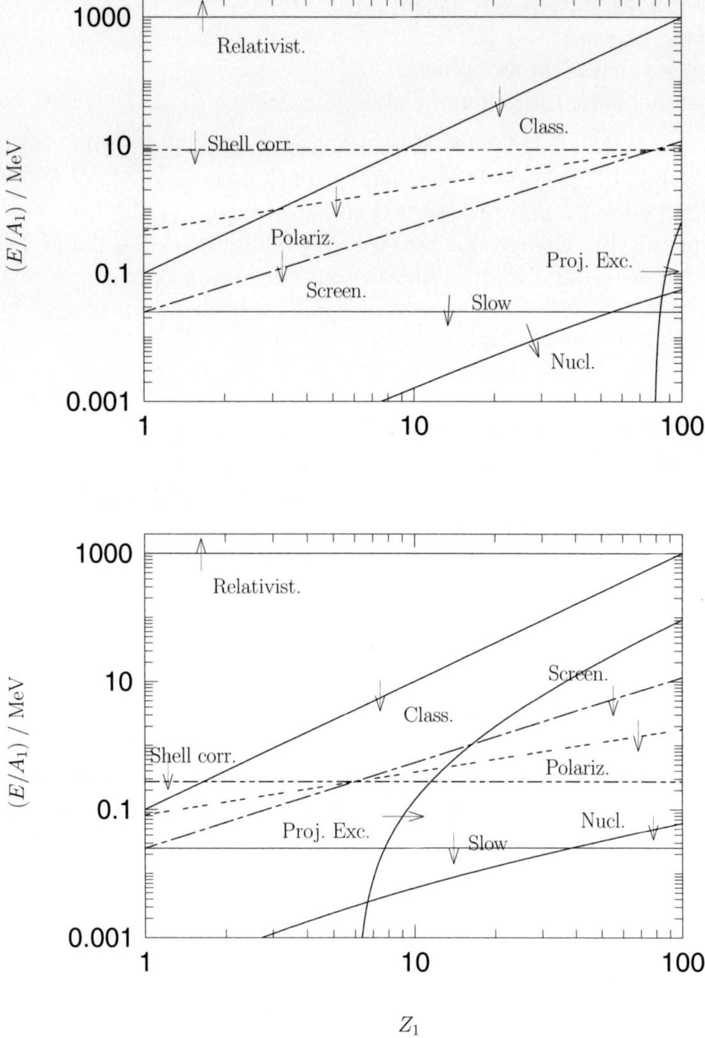

Fig. 4.2. Approximate limits between regimes of ion stopping. Arrows cross borderlines and point into a regime where the respective effect is dominating or significant. Upper graph: Gold target. Lower graph: Carbon target.

4.4 Regimes of Heavy-Ion Stopping

Figure 4.2 shows a qualitative survey of stopping regimes for a heavy (gold) and a light (carbon) target. Lines indicate rough limits between various regimes.

The thin horizontal line labelled 'Slow' at $E/A_1 = 25$ keV corresponding to $v = v_0$, the Bohr velocity, roughly delimits the regime of low-speed stopping where the ion speed is lower than the orbital velocities of all but the outermost target electrons.

The horizontal line labelled 'Shell corr.' marks the instance where the projectile speed equals the Thomas-Fermi velocity $Z_2^{2/3} v_0$ of the target electrons. Below this line the motion of target electrons cannot be ignored.

A third horizontal line labelled 'Relativist.' at $E/A_1 = 1$ GeV, close to the rest energy of the projectile, indicates the transition from the moderately-relativistic to the highly-relativistic velocity range.

According to Bohr (1948), a moving ion in charge equilibrium carries electrons with orbital velocities exceeding the projectile speed v. This defines a rough borderline labelled 'Screen.' at the Thomas-Fermi velocity $v = Z_1^{2/3} v_0$ of the projectile. Projectiles are expected to be stripped of the majority of their electrons at energies exceeding that limit.

Processes involving excitation or ionization of the projectile must become competitive whenever the number of electrons accompanying the projectile is comparable to or greater than that on the target atom. The lines labelled 'Proj. Exc.' indicate that this effect must be insignificant for gold except for the heaviest ions. Conversely, the effect must be expected to be noticeable for carbon except for the lightest projectile ions.

The line labelled 'Nucl.' marks the transition from dominating electronic to dominating nuclear stopping. It has been estimated very roughly on the basis of the Lindhard and Scharff (1961) formula for velocity-proportional electronic stopping and the so-called Nielsen formula (Lindhard & al., 1963) for energy-independent nuclear stopping, to be discussed in Sect. 10.4.

The two remaining lines, 'Class.' and 'Polariz.' express fundamental aspects of Coulomb excitation that deserve special attention and will be discussed in the following section.

4.5 Target Excitation

The dominating contribution to the stopping of light charged particles is excitation and ionization of target electrons ('electronic stopping') over a very wide energy/velocity range. The process is well described by the Bethe theory and its extensions (ICRU, 1993) which treat the Coulomb interaction between the projectile and the target electrons by quantal scattering theory in the first Born approximation. This results in a strict proportionality between the stopping force and the square of the projectile charge, Z_1^2. A lower limit for the range of validity of the first Born approximation is set by the requirement on the Sommerfeld parameter $Z_1 v_0/v$ to be < 1, or

$$v > Z_1 \alpha c, \tag{4.1}$$

where $\alpha = v_0/c \simeq 1/137$ is the fine structure constant, or

$$E/A_1 > Z_1^2 \times 25 \text{ keV}. \tag{4.2}$$

The Sommerfeld criterion is not a *necessary* criterion, i.e., nothing is said about the accuracy of the first Born approximation when the criterion is not fulfilled. It is roughly complementary to the Bohr criterion which delimits the range of validity of a description of Coulomb collisions in terms of classical orbits by the requirement that the Bohr parameter

$$\kappa = \frac{2Z_1 v_0}{v} > 1, \tag{4.3}$$

i.e.,

$$E/A_1 < Z_1^2 \times 100 \text{ keV}. \tag{4.4}$$

When this criterion is fulfilled, characteristic dimensions of a classical Kepler orbit exceed the de Broglie wavelength so that construction of a wave packet is possible which follows approximately the classical orbit. The Bohr criterion is a *sufficient* criterion.

The fact that there is an overlap regime between the two criteria promises a smooth transition between the classical and the Born regime. Only the line specified by (4.4), labelled 'classical', has been included in Fig. 4.2.

In addition to the Bohr parameter κ, a second dimensionless parameter can be constructed[1],

$$\frac{1}{\xi} = \frac{Z_1 v_0 \hbar \omega}{mv^3}, \tag{4.5}$$

which limits the range of validity of the Born approximation. Here ω is the frequency of a characteristic target resonance, e.g., $\omega = I/\hbar$ where I is the *mean excitation energy*, the 'I-value' of a target atom. Equation (4.5) determines the order of magnitude of the leading correction term in the Born series, the Barkas-Andersen effect that depends on the sign of the projectile charge and will be discussed in Sect. 5.4, where also an alternative interpretation of the effect is discussed which does not make reference to the Born series. This correction becomes substantial for

$$v \lesssim (Z_1 Z_2)^{1/3} v_0 \tag{4.6}$$

and is included under the label 'polarization'.

It is seen in Fig. 4.2 that for the lightest ions there is a wide energy range in which the uncorrected Bethe formula is valid, particularly for a light target.

---

[1] It is emphasized that despite the occurrence of $\hbar$, the parameter specified in (4.5) is a purely classical quantity since $v_0 \hbar \omega = e^2 \omega / 4\pi \epsilon_0$ is independent of Planck's constant. This notation – which does not distinguish between gaussian and SI units – is used in order to facilitate reference to the original literature that utilizes gaussian units.

Conversely, for the heaviest ions an even wider range of validity is expected for a classical model. However, the range of applicability of the pure Bohr model without screening and other corrections is quite narrow, in particular for light targets for which projectile excitation contributes.

The relative significance of screening, polarization and shell correction is reversed at $Z_1 = Z_2$. Going from high to low velocities, the shell correction comes first for $Z_1 < Z_2$ while screening does so for $Z_1 > Z_2$.

The main conclusion to be drawn from Fig. 4.2 is the fact that corrections to the simplest models (uncorrected Bethe and uncorrected Bohr formula) are required over a substantial portion of the parameter space, and that one single dominating correction is an exception rather than the rule. Taking due care of simultaneously acting effects in a nonlinear theory is a nontrivial task that, however, is facilitated if one takes the starting point in a classical-orbit description of the scattering process.

4.6 Interaction Range

Both the classical theory and the Born approximation predict the Coulomb interaction of a projectile moving with a speed v to be effectively limited to within the adiabatic radius

$$a = \frac{v}{\omega},\tag{4.7}$$

where ω is the resonance frequency of the respective electronic-excitation channel. This quantity is useful to consider in an assessment of the contribution to stopping from individual target shells as well as in estimates of the significance of collective effects.

In the classical regime the energy transfer to a target electron increases with decreasing impact parameter up to the limit defined by the conservation laws of energy and momentum. In the Born regime an effective upper limit is reached at an impact parameter of the order of the de Broglie wavelength. This has the consequence that hard electronic interactions are more important in the classical than in the Born regime. Taken together with (4.7) this implies that excitation of inner shells increases in significance with increasing Z_1 at constant v.

4.7 Role of Projectile Electrons

A distinct feature of heavy-ion stopping is the presence of electrons on the projectile at all but the highest velocities. Projectile screening due to these electrons tends to reduce the Coulomb interaction between the projectile and the target electrons, but at the same time projectile excitation may become a noticeable energy-loss channel. Moreover, the very processes of electron capture and loss cost energy, and fluctuations in charge state give rise to energy-loss straggling (*charge-exchange straggling*).

Projectile excitation and ionization have usually been treated by, roughly speeking, inverting the roles of target and projectile. This approach is problematic, and attention will be given to a revised procedure.

4.8 Nuclear Stopping

Energy loss to recoiling nuclei ('nuclear stopping')[2] is insignificant in the region of weakly-screened Coulomb interaction with target electrons but increases in relative significance as electronic excitation channels close with decreasing projectile speed. Electronic stopping depends on projectile *speed* while nuclear stopping depends on projectile *energy*. Therefore the point of crossover between nuclear and electronic stopping depends on Z_1 (and, less sensitively, Z_2), as is seen in Fig. 4.2.

References

Arista, N. R. (2002). "Energy loss of heavy ions in solids: non-linear calculations for slow and swift ions," Nucl. Instrum. Methods B **195**, 91–105.

Bethe, H. (1930). "Zur Theorie des Durchgangs schneller Korpuskularstrahlen durch Materie," Ann. Physik **5**, 324–400.

Bohr, N. (1940). "Scattering and stopping of fission fragments," Phys. Rev. **58**, 654–655.

Bohr, N. (1948). "The penetration of atomic particles through matter," Mat. Fys. Medd. Dan. Vid. Selsk. **18 no. 8**, 1–144.

Bohr, N. and Lindhard, J. (1954). "Electron capture and loss by heavy ions penetrating through matter," Mat. Fys. Medd. Dan. Vid. Selsk. **28 no. 7**, 1–31.

Brandt, W. and Kitagawa, M. (1982). "Effective stopping-power charges of swift ions in condensed matter," Phys. Rev. B **25**, 5631–5637.

Firsov, O. B. (1959). "A qualitative interpretation of the mean electron excitation energy in atomic collsions," Zh. Eksp. Teor. Fiz. **36**, 1517–1523, [English translation: Sov. Phys. JETP **9**, 1076-1080 (1959)].

Gombas, P. (1956). "Statistische Behandlung des Atoms," S. Flügge, ed., *Handbuch der Physik*, volume 36, 109–231 (Springer, Berlin).

Grande, P. L. and Schiwietz, G. (1998). "Impact-parameter dependence of the electronic energy loss of fast ions," Phys. Rev. A **58**, 3796–3801.

Grande, P. L. and Schiwietz, G. (2002). "The unitary convolution approximation for heavy ions," Nucl. Instrum. Methods B **195**, 55–63.

---

[2]In order to avoid confusion with processes on the nuclear scale, this atomic process has occasionally been denoted 'elastic stopping'. This notion is avoided here because the chance for confusion would become considerably greater, cf. footnote 1 on page 85.

ICRU (1984). *Stopping powers for electrons and positrons*, volume 37 (ICRU Report, International Commission of Radiation Units and Measurements, Bethesda, Maryland).

ICRU (1993). *Stopping powers and ranges for protons and alpha particles*, volume 49 (ICRU Report, International Commission of Radiation Units and Measurements, Bethesda, Maryland).

Knipp, J. and Teller, E. (1941). "On the Energy Loss of Heavy Ions," Phys. Rev. **59**, 659–669.

Lamb, W. E. (1940). "Passage of uranium fission fragments through matter," Phys. Rev. **58**, 696–702.

Lifschitz, A. F. and Arista, N. (1998). "Velocity-dependent screening in metals," Phys. Rev. A **57**, 200–207.

Lindhard, J. and Scharff, M. (1961). "Energy dissipation by ions in the keV region," Phys. Rev. **124**, 128–130.

Lindhard, J., Scharff, M. and Schiøtt, H. E. (1963). "Range concepts and heavy ion ranges," Mat. Fys. Medd. Dan. Vid. Selsk. **33 no. 14**, 1.

Maynard, G., Zwicknagel, G., Deutsch, C. and Katsonis, K. (2001). "Diffusion-transport cross section and stopping power of swift heavy ions - art. no. 052903," Phys. Rev. A **63**, 052903-1–14.

Northcliffe, L. C. (1963). "Passage of heavy ions through matter," Ann. Rev. Nucl. Sci. **13**, 67–102.

Paul, H. (2003). "Stopping power for light ions," URL `www.exphys.uni-linz.ac.at/stopping`.

Sigmund, P. and Schinner, A. (2000). "Binary stopping theory for swift heavy ions," Europ. Phys. J. D **12**, 425–434.

5 Electronic Stopping of Point Charges

5.1 Classical Theory

The prediction of the Bohr (1913, 1915) theory may be expressed in terms of the stopping number L defined by (3.1),

$$L = \sum_j f_j L_j(\xi_j, \beta) \tag{5.1}$$

with

$$L_j(\xi_j, \beta) = L_{\mathrm{Bohr}} = \ln(C\xi_j) - \ln\left(1 - \beta^2\right) - \frac{\beta^2}{2}, \qquad \left(\beta = \frac{v}{c}\right) \tag{5.2}$$

and

$$\xi_j = \frac{mv^3}{Z_1 v_0 \hbar \omega_j}, \tag{5.3}$$

i.e., the reciprocal of the expression given in (4.5), where ω_j and f_j are the resonance frequency and weight factor for the jth target resonance such that

$$\sum_j f_j = 1. \tag{5.4}$$

Moreover,

$$C = 2e^{-\gamma} = 1.1229 \tag{5.5}$$

where $\gamma = 0.5772$ is Euler's constant[1]. The logarithmic form of (5.2) originates in an asymptotic expansion for large values of ξ_j. For $\xi_j \lesssim 1$, when (5.2) turns negative, alternative expressions are available (Sigmund, 1996; Lindhard and Sørensen, 1996; Sigmund, 1997), the simplest of which can be extracted from Lindhard and Sørensen (1996),

$$L_j(\xi) = \frac{1}{2} \ln\left[1 + (C\xi)^2\right]. \tag{5.6}$$

Equation (5.2) ignores the intrinsic motion of target electrons. Repairing this defect necessitates shell corrections, the importance of which increases from outer to inner shells.

---

[1]Cf. footnote 2 on page 7

The shell correction to the Bohr theory has been determined by Sigmund (2000) via re-evaluation of the Bohr theory allowing for initial motion of the target electron. The main outcome from that work is a clear dominance of small-impact-parameter collisions in the shell correction, i.e., collisions in which the effect of binding on the energy transfer is insignificant. This implies that the computation of shell corrections reduces to a binary-collision problem where both collision partners are in motion initially. The appropriate transformation to a coordinate frame in which the target particle is at rest, is well known (Sigmund, 1982) and reads

$$L_j(v) = \left\langle \frac{\boldsymbol{v} \cdot (\boldsymbol{v} - \boldsymbol{v}_e)\boldsymbol{v}}{|\boldsymbol{v} - \boldsymbol{v}_e|^3} \, L_{0j}(|\boldsymbol{v} - \boldsymbol{v}_e|) \right\rangle_j \tag{5.7}$$

at all nonrelativistic velocities, where L_j and L_{0j} are the shell-corrected and uncorrected stopping numbers, respectively, for the jth shell.

Equation (5.7) has been extended to the relativistic regime by Tofterup (1983).

5.2 Quantal Perturbation Theory

Bethe stopping theory for point charges has been discussed extensively in Fano (1963); Inokuti (1971); ICRU (1984) and ICRU (1993). The focus here is on aspects specific to higher values of Z_1 as well as more recent issues.

Equation (5.1) remains valid, but (5.2) is replaced by the familiar form

$$L_j(B_j, \beta) = L_{\text{Bethe}} = \ln B_j - \ln\left(1 - \beta^2\right) - \beta^2 \tag{5.8}$$

with

$$B_j = \frac{2mv^2}{\hbar\omega_j}. \tag{5.9}$$

Figure 4.2 shows that the shell correction is the leading add-on, in particular so for high-Z_2 targets where also the Barkas-Andersen effect may become substantial.

The intrinsic motion of target electrons is taken fully into account in the Bethe theory but is ignored in the derivation of (5.8). Shell corrections to the Bethe theory have been discussed extensively in ICRU (1993). With a few exceptions (McGuire, 1983; Bichsel, 2002), existing evaluations have been based on model systems for the target. ICRU (1993) focused on hydrogenic wave functions whereas the electron-gas model (Lindhard and Winther, 1964) has been most frequently employed in the literature (Bonderup, 1967; Chu and Powers, 1972b; Ziegler & al., 1985). Alternative approaches include the harmonic-oscillator model (Sigmund and Haagerup, 1986) as well as the kinetic theory (Sigmund, 1982; Oddershede and Sabin, 1984).

A very efficient method for computing shell corrections in the Born regime, making use of the binary theory, has been explored by Sigmund and Schinner (2002b).

5.3 Bloch Theory

The theory of Bloch (1933b) connects Bohr's classical with Bethe's quantal perturbation theory. A transparent formulation of this theory has been presented by Lindhard and Sørensen (1996). The essential feature of this approach is the replacement of a classical impact parameter by quantized angular momentum. This does not affect distant collisions – where the Bethe theory reproduces Bohr's result – but delivers different results for close collisions when the Bohr parameter κ, (4.3), is not $\ll 1$.

In the nonrelativistic limit the Bloch term was determined by Lindhard and Sørensen (1996) on the basis of the transport cross section

$$\sigma^{(1)} = 4\pi \left(\frac{\hbar}{mv}\right)^2 \sum_{\ell=0}^{\infty} \sin^2\left(\delta_\ell - \delta_{\ell+1}\right), \tag{5.10}$$

where δ_ℓ is the ℓ'th phase shift for elastic binary scattering of a free target electron on a point projectile. $\sigma^{(1)}$ is related to the stopping cross section via

$$S = mv^2 \sigma^{(1)}. \tag{5.11}$$

Evaluation on the basis of free-Coulomb scattering of target electrons is meaningful because only the *difference* between (5.10) and the perturbation limit of the same relation is evaluated. Binding is central for distant interactions where the perturbation limit delivers the exact result.

This procedure leads to the Bloch correction

$$\Delta L_{\text{Bloch}} = \psi(1) - \text{Re}\,\psi\left(1 + i\frac{Z_1 v_0}{v}\right), \tag{5.12}$$

where ψ represents the logarithmic derivative of the gamma function, $\psi(\zeta) = \mathrm{d}\ln\Gamma(\zeta)/\mathrm{d}\zeta$, and Re denotes the real part. When added to the Bethe logarithm (5.8), the fulldrawn curve in Fig. 5.1 is obtained which approaches the Bohr logarithm at low velocities.

The Bloch correction vanishes in the limit of large velocities[2] and behaves like $\ln(Cv/Z_1 v_0)$ at low velocities. Therefore, Bloch's stopping formula may be rewritten in the form

$$L_{\text{Bloch}} = L_{\text{Bethe}} + \Delta L_{\text{Bloch}} = \ln\frac{Cmv^3}{Z_1 v_0 \hbar\omega} + \ln\frac{2Z_1 v_0}{v} - \text{Re}\,\psi\left(1 + i\frac{Z_1 v_0}{v}\right)$$

$$= L_{\text{Bohr}} + \Delta L_{\text{invBloch}}, \tag{5.13}$$

---

[2] The Bloch correction (5.12) reduces to a power law $\propto Z_1^2$ at high speed. This has led to the terminology of a Z_1^4 correction to the Bethe formula which is frequently identified with the Bloch correction. This is evidently justified when the Bloch correction is small but leads to absurd results in the opposite limit, as is seen from the line labelled 'Bethe + Z_1^4 in Fig. 5.1.

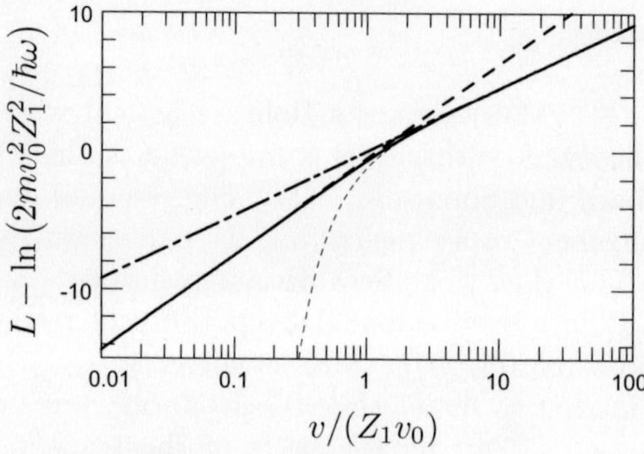

Fig. 5.1. Universal plot of simple Bohr and Bethe formulae for swift bare ions. Plotted is the stopping number L versus a scaled projectile speed. Shell, polarization and screening corrections neglected. Solid line: Bloch formula; dot-dashed and dashed lines: Bethe and Bohr logarithm L_{Bethe} and L_{Bohr} respectively; thin short-dashed line: Bethe logarithm plus high-speed approximation of Bloch correction ($\propto Z_1^4$). Figure from Sigmund (1997).

defining an *inverse-Bloch* correction $\Delta L_{\text{invBloch}}$ which vanishes at low speed but becomes substantial in the Bethe regime (de Ferrariis and Arista, 1984; Sigmund, 1996).

An accurate approximation to the Bloch stopping number was found by de Ferrariis and Arista (1984),

$$L_{\text{Bloch}} = L_{\text{Bethe}} + \Delta L_{\text{Bloch}} \simeq \ln \frac{2mv^2/\hbar\omega}{\sqrt{1 + (Z_1 v_0 \mathrm{e}^\gamma/v)^2}}$$

$$= \ln \frac{Cmv^3/Z_1 v_0 \hbar\omega}{\sqrt{1 + (Cv/2Z_1 v_0)^2}}. \quad (5.14)$$

Explicit evaluations of shell corrections to the Bloch formula are not known. However, since the Bloch correction (5.12) originates in close collisions, feasible shell corrections may be found by applying (5.7) which is exact for binary collisions.

While Bloch's calculation has not been extended to include terms of uneven order in Z_1, the magnitude of the Z_1^3 correction to the Bloch term can be estimated by comparison of calculations within classical and quantal perturbation theory. A Z_1^3 correction to the classical theory calculated for lithium on carbon (Schinner and Sigmund, 2000) was found to agree accurately with the corresponding result from quantal perturbation theory in the oscillator approach (Mikkelsen and Sigmund, 1989). To the extent that this finding may be generalized to other systems one may expect that a separate evaluation of Barkas-Andersen corrections to the Bloch term is of minor importance.

An extension of Bloch's theory to relativistic velocities was presented by Lindhard and Sørensen (1996) on the basis of the nonrelativistic approach sketched above, but employing relativistic scattering kinematics and the Dirac equation. This theory replaces earlier, more approximate treatments of corrections to the relativistic Bethe formula by Ahlen (1978, 1980) as well as the Mott correction (Jackson and McCarthy, 1972; Scheidenberger & al., 1994).

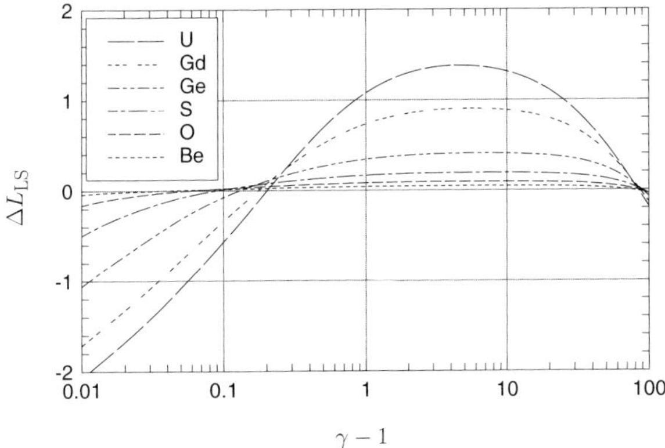

Fig. 5.2. Total relativistic correction according to Lindhard and Sørensen (1996), including correction for finite size of the projectile nucleus. The abscissa variable is $\gamma - 1 = 1/\sqrt{1 - v^2/c^2} - 1$.

The result may be summarized as a correction to the relativistic form of the Bethe stopping number,

$$\Delta L_{\mathrm{LS}} = \sum_{\varkappa=-\infty(\neq 0)}^{\infty} \left(\frac{|\varkappa|}{\eta^2} \frac{\varkappa - 1}{2\varkappa - 1} \sin^2\left(\delta_\varkappa - \delta_{\varkappa-1}\right) - \frac{1}{2|\varkappa|} \right)$$
$$+ \frac{1}{\eta^2} \sum_{\varkappa=1}^{\infty} \frac{\varkappa}{4\varkappa^2 - 1} \sin^2\left(\delta_\varkappa - \delta_{-\varkappa}\right) + \frac{v^2}{2c^2}, \quad (5.15)$$

where $\eta = Z_1 v_0/v = \kappa/2$, $\varkappa$ is an angular-momentum quantum number, and the phase shifts $\delta_\varkappa$ emerge from the Dirac equation.

This result is shown in Fig. 5.2 which represents a correction to the relativistic Bethe formula (5.8) including, if necessary, shell, screening, Barkas-Andersen and Fermi density-effect corrections. At very high velocities, deviations from pure Coulomb scattering in electron-nucleus scattering need to be considered ('finite-size effect') in the evaluation. Figure 5.3 indicates the magnitude of this correction for lead in aluminium. Good agreement with experimental data is found provided that also the Fermi density effect (ICRU, 1984) is included.

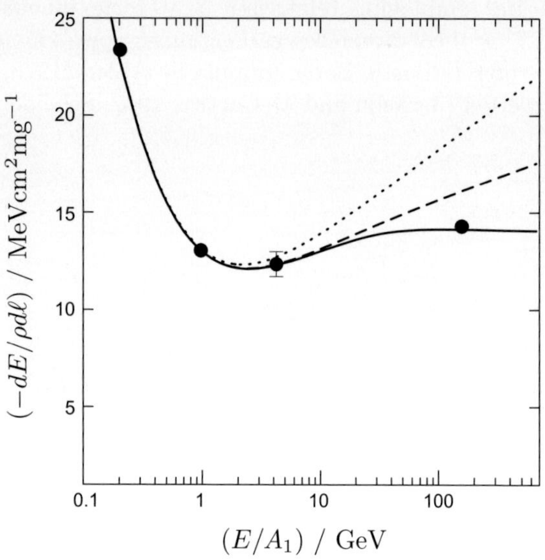

Fig. 5.3. Stopping of highly-relativistic ions: Experimental data for Pb in Al from Geissel and Scheidenberger (1998); Datz & al. (1996) and three theoretical curves based on Lindhard and Sørensen (1996). Dotted line: Point charge. Dashed line: Fermi density effect added. Solid line: Fermi density and nuclear-size effect added. From Geissel & al. (2002).

For ultrarelativistic projectiles, Lindhard and Sørensen (1996) derived the asymptotic expression

$$L \sim \ln \frac{1.62c}{R\omega_{\mathrm{p}}} \tag{5.16}$$

for the stopping number[3], where R is the nuclear radius and ω_{P} the plasma frequency reflecting the total electron density in the target.

Sørensen (2003) demonstrated that energy loss to bremsstrahlung and pair creation dominate over target excitation for $\gamma \gtrsim 10^3$ ($E \gtrsim 10^6$ MeV/u) in heavy materials.

---

[3] The value 1.62 is more accurate than the one given in the original paper (private communication by A. H. Sørensen).

5.4 Barkas-Andersen Effect

The Barkas effect denotes the difference in stopping between a particle and its antiparticle. The effect was discovered in the analysis of experiments aiming at meson masses (Smith *& al.*, 1953) and was ascribed to higher-order contributions to the Born series by Barkas *& al.* (1963) which cause deviations from the strict Z_1^2 dependence of the stopping force predicted by the Bethe theory, cf. (3.1) and (5.8).

The fact that this observation could become significant for ion stopping was not recognized until systematic high-precision measurements by Andersen *& al.* (1969) revealed that

- stopping forces on bare helium ions were *higher* than four times those on bare protons,
- that the difference increased with decreasing projectile speed and
- the effect could be quantified in terms of a term proportional to Z_1^3 contributing to the stopping force.

Even though the effect was small – at a level of several per cent for proton energies in the MeV range – it was these measurements that triggered several theoretical studies aiming at an understanding of this 'Z_1^3 effect'. The fact that the two types of measurement track the same physical phenomenon has led to the terminology of the *Barkas effect* (Lindhard, 1976). However, this terminology does not give justice to the seminal character of the measurements of Andersen *& al.* (1969) especially for heavy-ion stopping. Hence, the notion of the *'Barkas-Andersen effect'* is recommended here and in ICRU (2005).

The existence of a Z_1^3-proportional correction must evidently be the cause of serious concern in any theory of stopping for high-Z_1 ions. Indeed, the lack of a theory of the Barkas-Andersen effect going beyond the first correction term in a series valid for small Z_1 was a major obstacle for a long time toward a theory of heavy-ion stopping at intermediate velocities, as was mentioned in Ch. 1.

Ashley *& al.* (1972) presented a theoretical evaluation of the Z_1^3 correction within classical perturbation theory for large impact parameters. An equivalent quantal evaluation by Hill and Merzbacher (1974) confirmed their results. Those authors also proposed the notion of 'polarization effect'[4]. Both models left open the question of a possible higher-order correction from close collisions. That aspect was considered by Lindhard (1976) who analysed the Barkas-Andersen effect in terms of a deviation from pure Coulomb scattering instead of a deviation from first-order perturbation theory.

Ashley *& al.* (1972) derived an expression for the energy loss $w(p)$ versus impact parameter p in an individual interaction. While the validity of their

---

[4]The terms 'Barkas-Andersen correction' and 'polarization correction' will be used synonymously, while the term Z_1^3 correction will be reserved to the leading correction in Z_1 to the Bethe theory.

expression has been confirmed by all subsequent calculations, estimates of the Z_1^3-contribution to the *stopping cross section* have varied considerably, dependent on input. Ashley & al. (1972), asserting that there was no Z_1^3-term for close collisions, introduced a cutoff impact parameter. The choice of this parameter is crucial since $w(p)$ determined in this manner diverges strongly at small p. Jackson and McCarthy (1972) followed the same scheme although with a different choice of cutoff, and ICRU (1993) treats the cutoff as a fitting parameter to measured stopping forces. The estimate by Lindhard (1976) avoided the impact-parameter picture and led to an estimate of the Z_1^3 correction to the stopping cross section about twice the one proposed by Jackson and McCarthy (1972). Dimensional arguments based on characteristic length parameters demonstrated that the parameter (4.5) – which already emerged from the calculation of Ashley & al. (1972) – is the dominating characteristic of the Barkas-Andersen effect.

Complete evaluations of the Z_1^3 correction in the stopping cross section were based on the electron-gas model (Esbensen, 1976; Esbensen and Sigmund, 1990) and the harmonic-oscillator model (Mikkelsen and Sigmund, 1989), with a considerable followup literature.

According to (4.5) the Barkas-Andersen correction increases with decreasing speed. This has drastic consequences in antiproton stopping where the correction is negative and thus eventually gives rise to a change of sign in the stopping force. This marks the breakdown of a description of the Barkas-Andersen effect in terms of a series expansion in Z_1 and indicates the need for nonlinear stopping theory.

Much of the existing literature on this item – listed by Arista and Lifschitz (1999) – is based on the electron-gas model, makes reference to Echenique & al. (1981), and is geared toward the low-speed (velocity-proportional) regime. Approaches valid also at intermediate velocities were proposed by Mikkelsen and Flyvbjerg (1992) for the harmonic-oscillator model (although not applied in practice), by Schiwietz (1990) (applied mostly to H and He ions), by Lifschitz and Arista (1998), and by Sigmund and Schinner (2001).

Scaling properties of the Barkas-Andersen effect have been studied by Sigmund and Schinner (2003). It was found that for bare ions the Barkas ratio, i.e., the ratio L_+/L_- of the stopping numbers for an ion and its anti-ion, was almost independent of the atomic number Z_1 when plotted versus the Bohr variable $\xi = mv^3/Z_1 v_0 \hbar\omega$, cf. (5.3), while a more complex dependence was found on Z_2. For ions in charge equilibrium, the magnitude of the Barkas ratio was found to *decrease* with increasing atomic number Z_1 because of increasing screening at constant ξ.

Figure 5.4 shows a comparison of theoretical predictions with measurements on antiproton stopping in silicon. It is seen that while five theoretical predictions differ in details, the good overall agreement indicates that the Barkas-Andersen effect is well described at least for light ions.

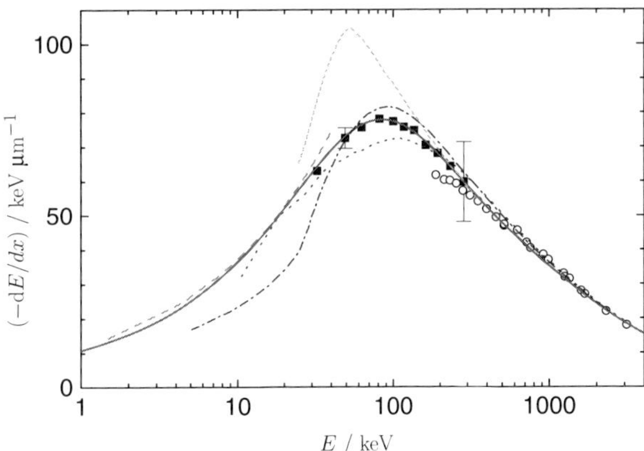

Fig. 5.4. Stopping of antiprotons in Si: Comparison of theoretical predictions by Sørensen (1990) (long-dashed line), Møller & al. (1997) (dotted line), Arista and Lifschitz (1999) (dot-dashed line), Arbó & al. (2000) (short-dashed line) and Sigmund and Schinner (2002b) (solid line) with experimental data from Andersen & al. (1989); Medenwaldt & al. (1991) (open circles) and Møller & al. (1997) (filled squares). From Sigmund and Schinner (2001).

5.5 Fermi Density Effect

It was seen in Fig. 5.3 that a Fermi density correction needs to be allowed for at extreme relativistic velocities. For a detailed exposition of this effect, reference is made to ICRU (1984).

5.6 *I*-Values and Oscillator-Strength Spectra

The excitation spectrum of the target as expressed by data sets (ω_j, f_j) is the main numerical input into both quantal and classical stopping formulae, as is seen from (5.1), (5.3) and (5.8). Within the range of validity of the logarithmic expressions (5.2) and (5.8), only the mean excitation energy I defined by

$$\ln I = \sum_j f_j \ln(\hbar\omega_j) \tag{5.17}$$

is of interest, but at lower projectile speed more detailed knowledge is required.

An extensive discussion of the determination of I-values was given in ICRU (1984) and ICRU (1993). I-values given there are mainly extracted from stopping measurements with protons.

Precision measurements on proton stopping have been performed in a velocity range where shell corrections are not negligible and where a Z_1^3 correction is found necessary. Therefore, I-values given in ICRU (1993) depend to a significant extent on theory available in 1984. Shortcomings of this procedure have become increasingly clear. Therefore, the ICRU has started preparations for a re-evaluation of key data including total I-values.

In the context of the present report, characterizing the excitation spectrum by a single I-value would be inadequate. Therefore, a closer inspection of available data on oscillator strengths was found appropriate.

In general the sum over f_j is replaced by an integral over a continuous spectrum of dipole oscillator strengths $f'(\omega)$ which is related to the long-wavelength dielectric function $\varepsilon(\omega)$ through

$$f'(\omega) = -\frac{2\epsilon_0 m}{\pi n_e e^2}\, \omega \operatorname{Im} \frac{1}{\varepsilon(\omega)} \tag{5.18}$$

where Im denotes the imaginary part,

$$\int_0^\infty \mathrm{d}\omega\, f'(\omega) = 1, \tag{5.19}$$

and $n_e = nZ_2$ is the number of electrons per volume. Since $\varepsilon(\omega)$ can be expressed by the complex refractive index $\mathsf{n}(\omega)+i\mathsf{k}(\omega)$, the oscillator strength spectrum may also be written in the form

$$f(\hbar\omega) = 1.5331 \cdot 10^{-3}\, \frac{A_2}{\rho}\, \frac{\hbar\omega\, \mathsf{nk}}{\left(\mathsf{n}^2 + \mathsf{k}^2\right)^2} \tag{5.20}$$

$\hbar\omega$ in eV; $f(\hbar\omega)$ in eV^{-1}

with the normalization

$$\int_0^\infty \mathrm{d}(\hbar\omega) f(\hbar\omega) = Z_2. \tag{5.21}$$

The function f differs from f' only by the normalization.

Oscillator-strength spectra have been determined theoretically on the basis of Slater orbitals (Dehmer & al., 1975). These spectra have been discretized into subshell frequencies and oscillator strengths and tabulated by Oddershede and Sabin (1984) for $1 \leq Z_2 \leq 36$.

Optical constants for numerous solids including covalent and ionic compounds have been tabulated over a wide frequency range (Palik, 1985, 1991, 1996, 2000). Equivalent information may be extracted from a compilation of x-ray scattering and absorption data (Henke & al., 1993). Similar data for atomic and molecular gases may be extracted from Berkowitz (1979, 2002).

A procedure employed to determine (f_j, ω_j) for determining stopping forces has been described by Sigmund and Schinner (2002a). The resulting data sets specify I-values, but those I-values do not enter directly as input.

Table 5.1. Selected *I*-values calculated from oscillator-strength spectra employed in ICRU (2005) compared with recommended values from ICRU (1993).

Element	Present	ICRU49	Element	Present	ICRU49
Be	64.7	63.7±3	Ge	399.1	350±11
C (amorphous)	86.0	81.0	Kr	390.6	352±26
N_2	78.5	82.0 ±2	Mo	424.0	424±15
O_2	94.6	95.0±2	Ag	447.5	470±10
Ne	135.5	137±4	Sn	457.8	488±15
Al	158.3	166±2	Xe	511.9	482
Si	169.5	173±3	W	719.0	727±30
Ar	182.5	188±10	Pt	751.6	790±30
Ti	241.8	233±5	Au	741.9	790±30
Fe	291.1	286±9	Pb		823±30
Ni	301.1	311±10	U		890±30
Cu	326.3	322±10			

Table 5.1 compares selected *I*-values so determined with values recommended in ICRU (1993). Quite good agreement is found in most cases where recommended values were based on measurements, taking into account the above reservations.

An alternative description of excitation spectra, applied frequently in stopping theory, is the dielectric theory by Lindhard and Scharff (1953), – often called local-density or local-plasma approximation – where the summation over frequencies is replaced by an integration over space such that

$$L = \frac{1}{Z_2} \int d^3 r n_e(\boldsymbol{r}) L\left(\frac{2mv^2}{\hbar\omega_P(\boldsymbol{r})}\right). \tag{5.22}$$

Here $n_e(\boldsymbol{r})$ is the electron density in a target atom ($\int d^3 r n_e(\boldsymbol{r}) = Z_2$),

$$\omega_P = \sqrt{\frac{n_e e^2}{\epsilon_0 m}} \tag{5.23}$$

the plasma frequency of a free-electron gas with density n_e, and $L(2mv^2/\hbar\omega_P)$ the stopping number evaluated for a free-electron gas including shell and (possibly) other corrections. The occurrence of the Bethe factor in the stopping number L is related to the fact that this picture has only been applied within quantal perturbation theory[5].

---

[5] In practice, (5.22) is usually (Bonderup, 1967; Chu and Powers, 1972b) evaluated by employing a shell correction expansion (Lindhard and Winther, 1964) of

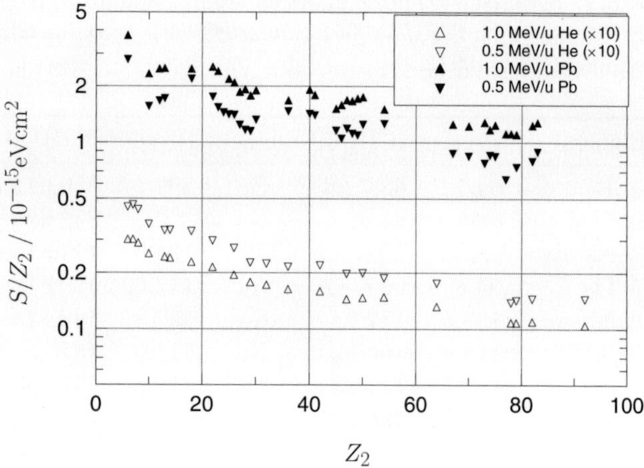

Fig. 5.5. Comparison of stopping cross sections per target electron for bombardment with Pb ions (closed symbols) (Geissel, 1982)) and He ions (open symbols) (ICRU, 1993). From Sigmund et al. (2003)

5.7 Z_2 Structure

According to the Thomas-Fermi estimate of the frequency mentioned in Sect. 4.3, the mean excitation energy should vary as $I \simeq Z_2 I_0$ with some universal constant I_0. This behavior, predicted by Bloch (1933a), is well confirmed as a first approximation, but superimposed on this monotonic increase is an oscillatory behavior as a function of Z_2, called Z_2 structure, which is well documented experimentally for protons and helium ions (ICRU, 1984, 1993), where it causes structure in the stopping cross section versus atomic number at constant speed. Both effects have been explained within the framework of dielectric theory (Chu and Powers, 1972a,b).

Inspection of (5.9) shows that within the range of validity of the Bethe theory the amplitude of observable oscillations in the stopping cross section must increase with decreasing speed. This has several reasons,

- The effect of a variation of $\hbar\omega_j$ with Z_2 becomes the more pronounced the smaller the numerator $2mv^2$,
- Such variations are most pronounced in outer target shells. Since inner-shell excitation channels close one by one with decreasing projectile speed, only those shells that produce the most pronounced oscillations contribute at low speed.

---

the stopping number in powers of v^{-2}. The first term in that expansion is the Bethe logarithm where a numerical factor $1/\chi$ is added to the argument with $\chi \simeq \sqrt{2}$, accounting approximately for atomic binding which is neglected in a Fermi-gas model (Lindhard and Scharff, 1953).

– Shell corrections tend to amplify Z_2 structure caused by the variation of ω_j with Z_2: A low value of ω_j is accompanied by a low orbital speed and hence by a low (negative) shell correction, and vice versa (Oddershede & al., 1983).

Below the classical limit, (5.2) replaces the Bethe logarithm and Z_2 structure tends to be enhanced further by the factor Z_1^{-1} under the Bohr logarithm. This is illustrated in Fig. 5.5 which compares stopping cross sections *per target electron* measured for Pb ions at 0.5 and 1.0 MeV/u with the corresponding values for He. The respective screening limits lie at 0.47 and 0.040 MeV/u. Even though the measurements with lead ions do not pertain to completely-stripped ions, Z_2 structure is clearly more pronounced than for He in this velocity range.

References

Ahlen, S. P. (1978). "Z_1^7 stopping-power formula for fast heavy ions," Phys. Rev. A **17**, 1236–1239.

Ahlen, S. P. (1980). "Theoretical and experimental aspects of the energy loss of relativistic heavily ionizing particles," Rev. Mod. Phys. **52**, 121–173.

Andersen, H. H., Simonsen, H. and Sørensen, H. (1969). "An experimental investigation of charge-dependent deviations from the Bethe stopping power formula," Nucl. Phys. **A125**, 171–175.

Andersen, L. H., Hvelplund, P., Knudsen, H., Møller, S. P., Pedersen, J. O. P., Uggerhøj, E., Elsener, K. and Morenzoni, E. (1989). "Measurement of the Z_1^3 contribution to the stopping power using MeV protons and antiprotons – the Barkas effect," Phys. Rev. Lett. **62**, 1731–1734.

Arbó, D. G., Gravielle, M. S. and Miraglia, J. E. (2000). "Second-order born collisional stopping of ions in a free-electron gas," Phys. Rev. A **62**, 032901–1–7.

Arista, N. R. and Lifschitz, A. F. (1999). "Nonlinear calculation of stopping powers for protons and antiprotons in solids: the Barkas effect," Phys. Rev. A **59**, 2719–2722.

Ashley, J. C., Ritchie, R. H. and Brandt, W. (1972). "Z_1^3 Effect in the stopping power of matter for charged particles," Phys. Rev. B **5**, 2393–2397.

Barkas, W. H., Dyer, J. N. and Heckman, H. H. (1963). "Resolution of the Σ^--mass anomaly," Phys. Rev. Lett. **11**, 26–28.

Berkowitz, J. (1979). *Photoabsorption, photoionization and photoelectron spectroscopy* (Academic Press, New York).

Berkowitz, J. (2002). *Atomic and molecular photoabsorption. Absolute total cross sections* (Academic Press, San Diego).

Bichsel, H. (2002). "Shell corrections in stopping powers," Phys. Rev. A **65**, 052709–1–11.

Bloch, F. (1933a). "Bremsvermögen von Atomen mit mehreren Elektronen," Z. Phys. **81**, 363–376.

Bloch, F. (1933b). "Zur Bremsung rasch bewegter Teilchen beim Durchgang durch Materie," Ann. Physik **16**, 285–320.

Bohr, N. (1913). "On the theory of the decrease of velocity of moving electrified particles on passing through matter," Philos. Mag. **25**, 10–31.

Bohr, N. (1915). "On the decrease of velocity of swiftly moving electrified particles in passing through matter." Philos. Mag. **30**, 581–612.

Bonderup, E. (1967). "Stopping of swift protons evaluated from statistical atomic model," Mat. Fys. Medd. Dan. Vid. Selsk. **35 no. 17**, 1–20.

Chu, W. K. and Powers, D. (1972a). "Calculations of mean excitation energy for all elements," Phys. Lett. **40A**, 23–24.

Chu, W. K. and Powers, D. (1972b). "On the Z_2 dependence of stopping cross sections for low energy alpha particles," Phys. Lett. A **38**, 267–268.

Datz, S., Krause, H. F., Vane, C. R., Knudsen, H., Grafström, P. and Schuch, R. H. (1996). "Effect of nuclear size on the stopping power of ultrarelativistic heavy ions," Phys. Rev. Lett. **77**, 2925–2928.

de Ferrariis, L. and Arista, N. R. (1984). "Classical and quantum-mechanical treatments of the energy loss of charged particles in dilute plasmas," Phys. Rev. A **29**, 2145–2159.

Dehmer, J. L., Inokuti, M. and Saxon, R. P. (1975). "Systematics of dipole oscillator-strength distributions for atoms of the first and second row," Phys. Rev. A **12**, 102–121.

Echenique, P. M., Nieminen, R. M. and Ritchie, R. H. (1981). "Density functional calculation of stopping power of an electron gas for slow ions," Sol. St. Comm. **37**, 779–781.

Esbensen, H. (1976). *Contributions to detailed perturbation theory for slowing-down of charged particles*, Ph.D. thesis, Aarhus University.

Esbensen, H. and Sigmund, P. (1990). "Barkas effect in a dense medium: stopping power and wake field," Annals of Physics **201**, 152–192.

Fano, U. (1963). "Penetration of protons, alpha particles, and mesons," Ann. Rev. Nucl. Sci. **13**, 1–66.

Geissel, H. (1982). "Untersuchungen zur Abbremsung von Schwerionen in Materie im Energiebereich von (0,5 – 10) MeV/U," GSI-Report **82-12**, 21–29.

Geissel, H. and Scheidenberger, C. (1998). "Slowing down of relativistic heavy ions and new applications," Nucl. Instrum. Methods B **136-8**, 114–124.

Geissel, H., Weick, H., Scheidenberger, C., Bimbot, R. and Gardès, D. (2002). "Experimental studies of heavy-ion slowing down in matter," Nucl. Instrum. Methods B **195**, 3–54.

Henke, B. L., Gullikson, E. M. and Davies, J. C. (1993). "X-ray interactions: photoabsorption, scattering, transmission, and reflection at E = 50-30,000 eV, Z = 1-92," At. Data & Nucl. Data Tab. **54**, 181–342.

Hill, K. W. and Merzbacher, E. (1974). "Polarization in distant Coulomb collisions of charged particles with atoms," Phys. Rev. A **9**, 156–165.

ICRU (1984). *Stopping powers for electrons and positrons*, volume 37 (ICRU Report, International Commission of Radiation Units and Measurements, Bethesda, Maryland).

ICRU (1993). *Stopping powers and ranges for protons and alpha particles*, volume 49 (ICRU Report, International Commission of Radiation Units and Measurements, Bethesda, Maryland).

ICRU (2005). "Stopping of Heavy Ions," J. ICRU to appear.

Inokuti, M. (1971). "Inelastic collisions of fast charged particles with atoms and molecules- the Bethe theory revisited," Rev. Mod. Phys. **43**, 297–347.

Jackson, J. D. and McCarthy, R. L. (1972). "Z^3 corrections to energy loss and range," Phys. Rev. B **6**, 4131–4141.

Lifschitz, A. F. and Arista, N. (1998). "Velocity-dependent screening in metals," Phys. Rev. A **57**, 200–207.

Lindhard, J. (1976). "The Barkas effect – or Z_1^3, Z_1^4-corrections to stopping of swift charged particles," Nucl. Instrum. Methods **132**, 1–5.

Lindhard, J. and Scharff, M. (1953). "Energy loss in matter by fast particles of low charge," Mat. Fys. Medd. Dan. Vid. Selsk. **27 no. 15**, 1–31.

Lindhard, J. and Sørensen, A. H. (1996). "On the relativistic theory of stopping of heavy ions," Phys. Rev. A **53**, 2443–2456.

Lindhard, J. and Winther, A. (1964). "Stopping power of electron gas and equipartition rule," Mat. Fys. Medd. Dan. Vid. Selsk. **34 no. 4**, 1–22.

McGuire, E. J. (1983). "Extraction of shell corrections from Born-approximation stopping-power calculations in Al," Phys. Rev. A **28**, 49–52.

Medenwaldt, R., Møller, S. P., Uggerhøj, E., Worm, T., Hvelplund, P., Knudsen, H., Elsener, K. and Morenzoni, E. (1991). "Measurement of the stopping power of silicon for antiprotons between 0. 2 and 3 MeV," Nucl. Instrum. Methods B **58**, 1–5.

Mikkelsen, H. H. and Flyvbjerg, H. (1992). "Exact stopping cross section of the quantum harmonic oscillator for a penetrating point charge of arbitrary strength," Phys. Rev. A **45**, 3025–3031.

Mikkelsen, H. H. and Sigmund, P. (1989). "Barkas effect in electronic stopping power: rigorous evaluation for the harmonic oscillator," Phys. Rev. A **40**, 101–116.

Møller, S. P., Uggerhøj, E., Bluhme, H., Knudsen, H., Mikkelsen, U., Paludan, K. and Morenzoni, E. (1997). "Direct measurement of the stopping power for antiprotons of light and heavy targets," Phys. Rev. A **56**, 2930–2939.

Oddershede, J. and Sabin, J. R. (1984). "Orbital and whole-atom proton stopping power and shell corrections for atoms with Z ¡ 36," At. Data Nucl. Data Tab. **31**, 275–297.

Oddershede, J., Sabin, J. R. and Sigmund, P. (1983). "Predicted Z_2-structure and gas-solid difference in low-velocity stopping power of light ions," Phys. Rev. Lett. **51**, 1332–1335.

Palik, E. D. (1985). *Handbook of optical constants*, volume 1 of *Academic Press Handbook Series* (Academic Press, Orlando).

Palik, E. D. (1991). *Handbook of optical constants of solids*, volume 2 (Academic Press, Boston).

Palik, E. D. (1996). *Handbook of optical constants of solids*, volume 3 (Academic Press, Boston).

Palik, E. D. (2000). *Electronic handbook of optical constants of solids – version 1.0* (SciVision – Academic Press).

Scheidenberger, C., Geissel, H., Mikkelsen, H. H., Nickel, F., Brohm, T., Folger, H., Irnich, H., Magel, A., Mohar, M. F., Münzenberg, G., Pfützner, M., Roeckl, E., Schall, I., Schardt, D., Schmidt, K. H., Schwab, W., Steiner, M., Stöhlker, T., Sümmerer, K., Vieira, D. J., Voss, B. and Weber, M. (1994). "Direct observation of systematic deviations from the Bethe stopping theory for relativistic heavy ions," Phys. Rev. Lett. **73**, 50–53.

Schinner, A. and Sigmund, P. (2000). "Polarization effect in stopping of swift partially screened heavy ions: perturbative theory," Nucl. Instrum. Methods B **164-165**, 220–229.

Schiwietz, G. (1990). "Coupled-channel calculation of stopping powers for intermediate-energy light ions penetrating atomic H and He targets," Phys. Rev. A **42**, 296–306.

Sigmund, P. (1982). "Kinetic theory of particle stopping in a medium with internal motion," Phys. Rev. A **26**, 2497–2517.

Sigmund, P. (1996). "Low-velocity limit of Bohr's stopping-power formula," Phys. Rev. A **54**, 3113–3117.

Sigmund, P. (1997). "Charge-dependent electronic stopping of swift nonrelativistic heavy ions," Phys. Rev. A **56**, 3781–3793.

Sigmund, P. (2000). "Shell correction in Bohr stopping theory," Europ. Phys. J. D **12**, 111–116.

Sigmund, P., Fettouhi, A. and Schimmer, A. (2003). "Matherial dependence of electronic stopping," Nucl. Instrum. Methods B **209**, 19–25.

Sigmund, P. and Haagerup, U. (1986). "Bethe stopping theory for a harmonic oscillator and Bohr's oscillator model of atomic stopping," Phys. Rev. A **34**, 892–910.

Sigmund, P. and Schinner, A. (2001). "Binary theory of antiproton stopping," Europ. Phys. J. D **15**, 165–172.

Sigmund, P. and Schinner, A. (2002a). "Binary theory of electronic stopping," Nucl. Instrum. Methods B **195**, 64–90.

Sigmund, P. and Schinner, A. (2002b). "Binary theory of light-ion stopping," Nucl. Instrum. Methods B **193**, 49–55.

Sigmund, P. and Schinner, A. (2003). "Anatomy of the Barkas effect," Nucl. Instrum. Methods B **212**, 110 – 117.

Smith, F. M., Birnbaum, W. and Barkas, W. H. (1953). "Measurements of meson masses and related quantities," Phys. Rev. **91**, 765–766.

Sørensen, A. H. (1990). "Barkas effect at low velocities," Nucl. Instrum. Methods B **48**, 10–13.

Sørensen, A. H. (2003). "Stopping of relativistic ions; the pair production and bremsstrahlung channels," AIP conference proceedings in press.

Tofterup, A. L. (1983). "Relativistic binary-encounter and stopping theory: general expressions," J. Phys. B **16**, 2997–3003.

Ziegler, J. F., Biersack, J. P. and Littmark, U. (1985). "The stopping and range of ions in solids," J. F. Ziegler, ed., *The Stopping and Ranges of Ions in Matter*, volume 1 of *The Stopping and Ranges of Ions in Matter*, 1–319 (Pergamon, New York).

6 Electronic Stopping of Dressed Ions

6.1 Equilibrium and Non-Equilibrium Stopping

Figure 4.2 indicates that screening by electrons accompanying the projectile increases in importance with increasing Z_1. The same statement applies to projectile excitation. These, as well as related effects, are charge-state dependent.

The charge state of the projectile is well-defined and readily measurable as long as the target material is a gas, while the matter is more delicate for a solid stopping material (Betz, 1972). Charge states of ions penetrating solids are typically measured *after emergence*, and most often ions are allowed to travel over macroscopic distances before detection, with the possibility of electron loss by Auger emission . This problem is avoided in measurements of the charge state *during emergence* (Brunelle & al., 1997), but the experimental method involved, making use of hydrogen emission from the target surface and discussed by Geissel & al. (2002) is indirect and not yet understood in detail.

Theoretically the specification of a charge state for an ion moving in a dense medium requires a clear distinction between electrons moving with the projectile and electrons that do not. One complication in this context is the existence of convoy (or cusp) electrons which are emitted downstream with a velocity close to that of the emerging projectile. An operative definition of a charge state is possible in principle by measurement of x-ray satellites (Knudson & al., 1974), but this is not part of the standard routine.

Of primary interest is equilibrium stopping, i.e., stopping under charge-state equilibrium. This is a dynamic equilibrium characterized by probabilities (*charge fractions*) $P(v, q_1)$ for an ion to have a charge $q_1 e$ at speed v. One may then define an average equilibrium charge

$$\langle q_1 \rangle = \sum_{q_1} P(v, q_1) q_1 \tag{6.1}$$

and an equilibrium stopping cross section

$$\langle S \rangle = \sum_{q_1} P(v, q_1) S(v, q_1), \tag{6.2}$$

where the *frozen-charge stopping cross section* $S(v, q_1)$ characterizes the average stopping between two charge-changing events[1]. Energy losses due to charge exchange need to be added to (6.2) when significant. $S(v, q_1)$ may be determined experimentally under pre-equilibrium conditions, i.e., in measurements of the energy loss in thin targets, differential in entrance and/or exit charge state.

It is well established experimentally that equilibrium charge states of swift heavy ions are higher in solids than in gases (Lassen, 1951a,b). This has been ascribed to a density effect by Bohr and Lindhard (1954): Free-flight times are too short to allow excited ions to decay into their ground states. This results in enhanced electron-loss rates and thus in a shifted charge equilibrium. Although the matter has been the subject of debate (Betz and Grodzins, 1970), increasing evidence appears to support this model (Maynard & al., 2000).

Equilibrium charge states for heavy ions have been studied intensely. Results for solids have been summarized by an analytical formula for the mean charge measured after emergence (Shima & al., 1982)[2] as well as a comprehensive tabulation for amorphous carbon (Shima & al., 1992). A simple Thomas-Fermi estimate

$$\langle q_1 \rangle = Z_1 \left(1 - e^{-v/Z_1^{2/3} v_0} \right) \tag{6.3}$$

tends to represent the same data very well for projectiles at least up to argon. An example is given in Fig. 6.1. Following Nikolaev and Dmitriev (1968), Shima et al. prefer a scaling variable $\propto Z_1^{0.45} v_0$. More involved scaling procedures, allowing for gas-solid differences, have been examined by Schiwietz and Grande (2001).

6.2 Screening and Antiscreening

The influence on target excitation of electrons accompanying the projectile may be roughly classified into screening and antiscreening. Screening denotes the electrostatic effect of the reduction in electric field strength. This depends on the distance from the projectile nucleus and tends to reduce the stopping force. The concept of antiscreening arises from viewing the projectile as an aggregate of particles, each of which interacting individually with the target electrons, thus giving rise to enhanced stopping.

---

[1]Equation (6.2) remains valid also when $P(v, q_1)$ describes a non-equilibrium charge distribution.

[2]The analytical formula of Shima & al. (1982) has both explicit and implicit limitations. The authors state its range of validity as $Z_1 \geq 8$ and $4 \leq Z_2 \leq 79$. Cases have been found where $\langle q_1 \rangle$ slightly exceeds its maximum value Z_1 for heavy ions like iodine.

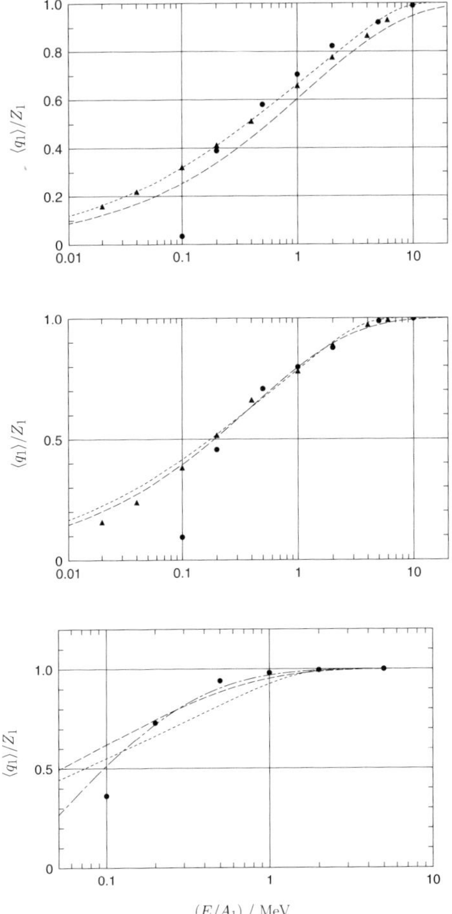

Fig. 6.1. Comparison between mean equilibrium charge states for Ar, O and Li in Si (top to bottom) predicted by (6.3) (dashed lines), interpolation formula by Shima *&* al. (1982) (dotted lines), tabulation by Shima *&* al. (1992) (triangles), measurements by Itoh *&* al. (1999) (dot-dashed line) and prediction from ETACHA code by Rozet *&* al. (1996) (circles).

Antiscreening is a relevant concept for projectile electrons with orbital radii that are not small compared to characteristic interaction distances. A noticeable effect has been found for neutral hydrogen (Kabachnik, 1993). The effect is intimately connected to electron loss and/or projectile excitation.

The significance of the effect decreases rapidly with increasing Z_1 (Sigmund, 1997).

6.3 Effective Charge and Quantal Perturbation Theory

Initially the effect of screening was incorporated into stopping theory via introduction of an effective charge which was thought to be close to the equilibrium ion charge (Bohr, 1940, 1941; Knipp and Teller, 1941). Later on, Northcliffe (1960, 1963) *defined* an effective charge

$$q_{1,\mathrm{eff}}e = \gamma Z_1 e \qquad (6.4)$$

via the ratio between the stopping force on a heavy ion and that on a proton at the same speed. Traditionally the effective-charge fraction γ^2 is determined empirically[3] by comparison of measured equilibrium stopping forces with stopping forces on protons or alpha particles, the latter originally being considered as point charges. Attempts of theoretical support to this concept were made by Yarlagadda & al. (1978), Brandt and Kitagawa (1982) and others.

It is recalled from Fig. 4.2 that the regime of significant screening lies entirely within the classical regime for both light and heavy targets. Thus, treating screening on the basis of quantal perturbation theory necessitates application of the Bloch correction. However, the Bloch correction has been evaluated only for bare ions[4]. In the absence of an evaluation for substantial screening, the atomic number Z_1 in the Bloch correction has usually been identified with the effective ion charge $\langle q_1 \rangle$. Arguments in favor of this choice have been brought forward (Northcliffe, 1960; Arista, 2002). However, the Bloch correction originates in close collisions. Hence, replacing Z_1 by γZ_1 is likely to to understimate the Bloch correction.

The above problem is avoided in a classical treatment: Within the classical regime, screening can be incorporated via a suitably chosen interaction potential, and the inverse-Bloch correction ensuring a smooth transition into the Born regime becomes substantial only at velocities were screening is of minor significance.

Figure 6.2 shows the effective-charge ratio γ^2 of oxygen in carbon, determined theoretically from the ratio of stopping numbers for O-C and He-C, both calculated for charge equlibrium according to (6.3). Also included is the corresponding curve for completely stripped oxygen and helium ions. If γZ_1 were an effective charge, the latter curve (dashed line) would have to be =

---

[3]Concerning the *effective-charge ratio* γ^2 as defined by (6.4) cf. footnote 2 on page 7.

[4]An exception is a study by Sørensen (2002) – utilized by Weick & al. (2002) – of the influence of weak screening (one electron per projectile ion) on the Lindhard-Sørensen term in the relativistic regime.

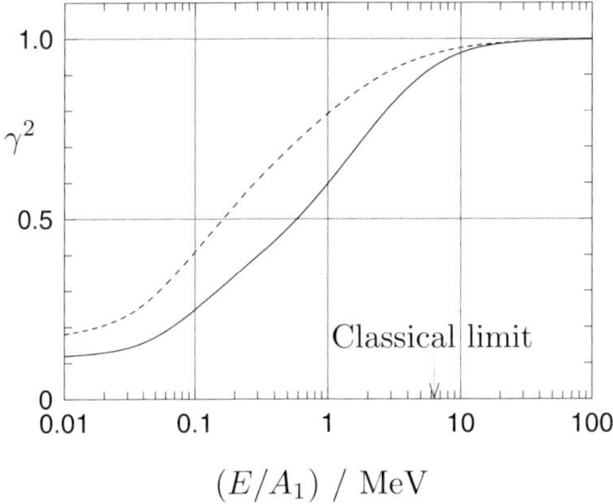

$$(E/A_1) \; / \; \text{MeV}$$

Fig. 6.2. Effective-charge ratio for oxygen in carbon, defined as $\gamma^2 = L_{\text{O}-\text{C}}/L_{\text{He}-\text{C}}$, calculated from binary theory. Solid curve: Charge equilibirum assumed for both oxygen and helium; dashed curve: Both oxygen and helium assumed to be completely stripped. From Sigmund and Schinner (2001a).

1 at all velocities. The fact that the curve falls off to about 0.2 reflects the transition from the Bethe to the Bohr regime which lies at a higher speed for oxygen than for helium. The point corresponding to $\kappa = 1$ for oxygen has been marked in the graph ('classical limit').

The topic has been discussed in considerable detail by Sigmund and Schinner (2001a). One may conclude from their discussion

- that the velocity dependence of the effective charge does not reflect that of the ion charge and
- that there is no reason to expect scaling properties in Z_1 and Z_2 similar to those of the ion charge.

These considerations indicate that there is no theoretical basis for scaling relations involving relative stopping forces. Further objections against the effective-charge postulate will be discussed in connection with Z_2 structure in section 8.7. Empirical findings are discussed in Chap. 4 of ICRU (2005).

6.4 Screened Potential

Brandt and Kitagawa (1982) established an explicit connection between ion charge and stopping force by invoking a potential

$$V(r) = -\frac{q_1 e^2}{r} - \frac{(Z_1 - q_1)e^2}{r} \, e^{-r/a_{\text{sc}}} \tag{6.5}$$

in a quantal perturbation theory for a free electron gas within the Lindhard (1954) scheme. The screening radius a_{sc} was determined on the basis of a modified Thomas-Fermi model. No Bloch correction was considered.

Equation (6.5), in combination with (5.22) and the effective-charge model described above, form the basis of the tabulation of Ziegler & al. (1985), where a_{sc} has been treated as a fitting parameter.

Similar schemes but involving more general screening functions have been explored (Kaneko, 1999; Grande and Schiwietz, 2002; Arista, 2002).

6.5 Classical Perturbation Theory

Screening was incorporated into Bohr's classical theory via (6.5) (Sigmund, 1997) but with a charge-dependent screening radius

$$a_{\mathrm{sc}} = 0.8853 a_0 Z_1^{-1/3} \left(1 - \frac{q_1}{Z_1}\right), \tag{6.6}$$

based on the model of Fermi and Amaldi (1934) of atomic ions. More general screening functions have been applied in classical theory by Maynard & al. (1996, 2001).

The combined effect of projectile screening and a polarization correction was considered by Schinner and Sigmund (2000) in a classically-based perturbation approach. The occurrence of unrealistically high Z_1^3 corrections (up to 100 %) indicated the need for an alternative, nonperturbative approach. This led to the binary theory (Sigmund and Schinner, 2000) which is to be described in more detail below.

6.6 Charge-Dependent Stopping

Several theoretical schemes to be discussed in the following deliver *partial stopping cross sections* or *frozen-charge stopping cross sections*, $S(v, q_1)$, which can be utilized to estimate the energy loss of a beam characterized by charge fractions $P(v, q_1)$ according to (6.2). Partial stopping cross sections have also been measured, and comparisons with theory have been made by Sigmund and Schinner (2001b), Grande and Schiwietz (2002), and in ICRU (2005). The statistical description of energy-loss spectra in the presence of charge exchange will be described in Sect. 13.

A question of practical significance is to what extent the equilibrium stopping cross section, (6.2), may be replaced by the stopping cross section taken at the equilibrium charge (6.1), i.e.,

$$\langle S(q_1) \rangle \overset{?}{\simeq} S(\langle q_1 \rangle). \tag{6.7}$$

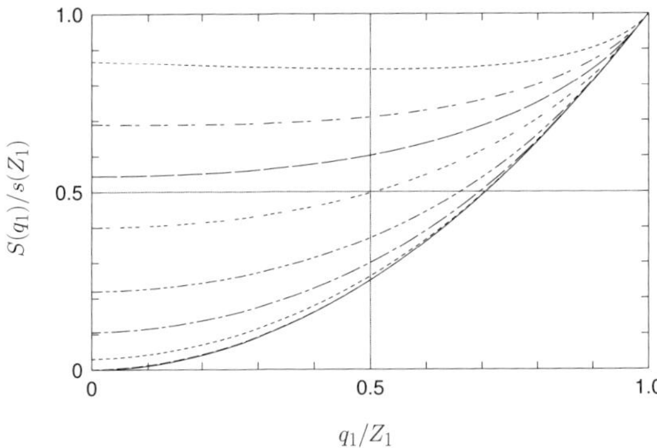

Fig. 6.3. Calculated ratio of stopping cross sections for frozen charge q_1 and bare ion, versus charge fraction. Curves for values 0.2, 0.5, 1, 2, 10, 20, 50 (top to bottom) of the parameter $2(Z_1/Z_2)^{2/3}$ (see text). From Sigmund (1997).

Figure 6.3 shows an estimate of stopping cross section versus ion charge, based on modified Bohr theory allowing for screening (Sigmund, 1997) but disregarding shell and Barkas-Andersen correction as well as projectile excitation. Curves are shown for a sequence of values of the parameter

$$s = \left(\frac{Z_1 m v_0^2}{I}\right)^{2/3} \simeq 2 \left(\frac{Z_1}{Z_2}\right)^{2/3}, \tag{6.8}$$

which was found to characterize the importance of screening[5]. For perfect screening, all curves would coincide with the parabola $(q_1/Z_1)^2$, the lowermost (thin full-drawn) curve in the figure. It is seen that such a behavior is only found for $Z_1 \gg Z_2$. For all other ion-target combinations the dependence of the partial stopping cross section on the instantaneous ion charge is weaker than the q_1^2 dependence that would be expected from complete screening in conjunction with the Bethe formula, and for $Z_1 \ll Z_2$ only a weak charge dependence is found.

Figure 6.4 shows related information predicted by the binary theory. Only curves for neutral projectiles are shown, as a function of beam energy. Qualitative results are very similar: For $Z_1 \gg Z_2$ stopping is very much reduced for the neutral projectile, except at the high-energy end where collision diameters get below the screening radius. Conversely, with decreasing ratio $Z_1 \leq Z_2$ the reduction of the stopping force by screening decreases monotonically. On the other hand, even the pronounced decrease for $Z_1 \gg Z_2$ will be counteracted to some extent by projectile excitation at low velocities. Thus, in most cases

---

[5]The last part of this relation implies the Bloch relation $I \propto Z_2$.

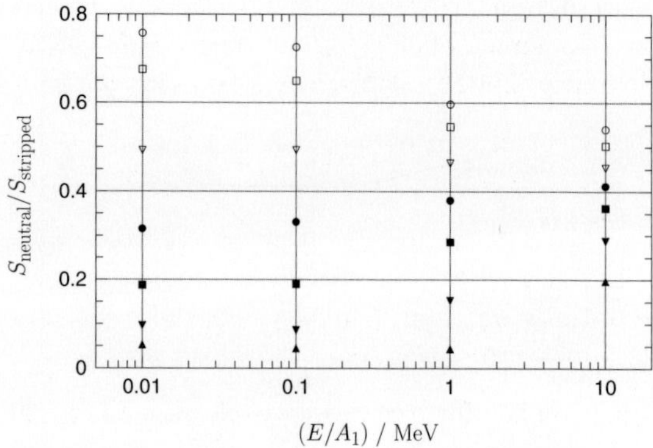

Fig. 6.4. Reduction factor in stopping from stripped to neutral projectile versus beam energy. Points for values 0.2, 0.5, 1, 2, 10, 20, 50 (top to bottom) of the parameter $2(Z_1/Z_2)^{2/3}$ (see text). Calculated from binary theory.

the overall dependence of the stopping cross section on q_1 must be less than quadratic. Based on these arguments one may give an upper bound on the error,

$$\frac{\langle S \rangle - S(\langle q \rangle)}{S(\langle q \rangle)} < \frac{\langle q_1^2 \rangle - \langle q_1 \rangle^2}{\langle q_1 \rangle^2}. \tag{6.9}$$

The quantity on the right-hand side may be extracted from tabulations by Shima & al. (1992). In general terms it is large for low-Z_1 ions – where only few charge states are involved – and falls off rapidly with increasing Z_1.

6.7 Projectile Excitation and Charge Exchange

Very roughly spoken, projectile excitation can be treated by inverting the roles of target and projectile. However, both practical and conceptual problems arise here that need careful consideration. Special complications arise for projectile ionization:

Consider first differences between target and projectile excitation,

- The projectile is charged. This implies that atomic data are needed not only for neutral atoms but also for a wide variety of ionic states.
- Whether neutral or charged, the projectile need not be in its ground state.
- Little systematics is known about the state of a penetrating ion or associated oscillator strengths.

These problems were addressed in terms of simple estimates by Sigmund and Schinner (2002a). Uncertainties associated with the state of the projectile were found to be small in practice, mostly because projectile excitation is a small effect except at velocities near and below v_0 (Sigmund and Schinner, 2002b). Moreover, two options were considered,

1. the number of electrons in every shell is reduced by a common factor q_1/Z_1, or
2. projectile states are filled up from the bottom,

although shifts in the excitation frequencies were ignored. The difference between the two options was found small in general.

For projectile ionization special considerations apply.

- Projectile ionization is a form of charge exchange. In equilibrium stopping the number of electron-loss events is equal to the number of electron-capture events in the average. Including only one of them in the formalism is not meaningful in general.
- The energy loss in an event leading to projectile ionization does not necessarily exceed that of projectile excitation because the speed of an ejected electron in the laboratory frame of reference may well be lower than that of the projectile.

As pointed out by Sigmund and Glazov (2003), the second item implies that energy loss due to projectile excitation has, when considered at all, been overestimated in the theoretical literature.

A further problem arises from the fact that an ion emerging from a penetrated layer in a charge state different from the incident one has changed its internal energy. Hence, the value of the measured energy loss depends on the chosen technique (Sigmund and Glazov, 1998): Uncritical analysis of energy losses measured by time-of-flight, magnetic or electrostatic analysis, or energy-dispersive detectors may provide different numerical values when converted to energy loss. This is significant in measurements of energy loss as a function of incident and/or exit charge.

Glazov (2002) presented a thorough theoretical analysis, based on the Born approximation, of projectile excitation. This continues earlier work by Kim and Cheng (1980) with the important difference that ionization is explicitly omitted. This complicates the analysis because of missing sum rules.

Electron capture is a more delicate issue, not only with regard to pertinent cross sections but also the definition of energy loss. Here, nonradiative electron capture is considered which is the dominating process at nonrelativistic velocities.

For a fast ion – with a speed substantially above the initial orbital speed of a captured electron – one may approximate the energy loss w_{capt} in a capture event by

$$w_{\mathrm{capt}} \simeq U_{\mathrm{init}} - U_{\mathrm{capt}} + \frac{mv^2}{2}, \tag{6.10}$$

where U_{init} is the ionization energy of the bound target electron, U_{capt} the ionization energy of the state into which capture takes place, and $mv^2/2$ the energy needed to accelerate the captured electron to the projectile speed v.

Electrons are typically captured into a high-angular-momentum state, i.e., an excited state of the projectile. In a gas target, such a state may decay into the ground state radiatively or via Auger emission. These processes do not lead to significant momentum changes of the projectile and, therefore, have to be omitted from (6.10).

For high-speed projectiles the dominating contribution in (6.10) is the last term. At the same time, cross sections for electron capture decrease rapidly with increasing speed. Therefore, equilibrium stopping forces are typically unaffected by capture. On the other hand, $mv^2/2$ may well exceed the average energy loss of a frozen charge in a foil and hence cause the energy-loss spectrum to be split into discrete portions reflecting the number of capture events encountered over a trajectory. Examples will be discussed in Sect. 12.3. If no distinction is made between target electrons excited into the continuum or a bound projectile state, energy loss to electron capture may be considered as being included in target excitation/ionization.

Cross sections for electron capture in the nonrelativistic as well as the relativistic velocity range have been tabulated by Ichihara & al. (1993). For comparisons with experimental results cf. Geissel & al. (2002).

6.8 Z_2 Structure

It was observed in Sect. 5.7 that for bare ions Z_2 structure gets increasingly pronounced with decreasing beam energy. For dressed ions Z_2 structure is also influenced by projectile screening. To appreciate this, recapitulate that Z_2 structure is generated primarily by variations of the outer-shell frequency ω_j with Z_2, which enters via the adiabatic radius v/ω_j: If ω_j is small, the effective interaction range and hence the stopping cross section is large, and vice versa. Now, in the presence of screening also the screening radius a_{sc} of the interaction needs to be considered. Screening is only effective for $a_{\mathrm{sc}} < v/\omega_j$, and, if so, the interaction range and hence the stopping force is increasingly determined by a_{sc}, which is only weakly dependent on the medium and hence tends to suppress Z_2 structure.

Figure 6.5 shows these trends for argon and proton bombardment. In either case structure increases with decreasing energy but seems to saturate below 100 keV/u. At the same time, structure is considerably more pronounced for proton than for argon bombardment. This has the implication that Z_2 structure is more pronounced in the effective-charge ratio $\gamma^2 = S(\mathrm{Ar})/S(\mathrm{H})$ than in the stopping cross section $S(\mathrm{Ar})$ itself. This is rather surprising, as the original motivation for introducing the effective-charge concept was the assumption that the effective charge was independent of or only weakly dependent on the stopping material.

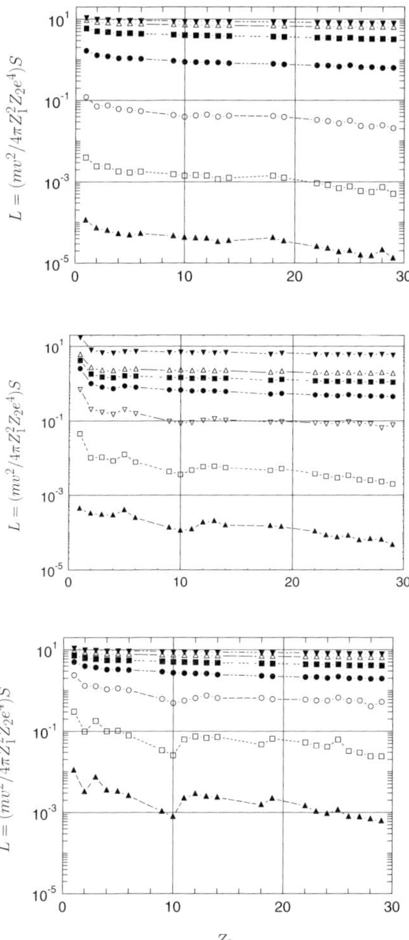

Fig. 6.5. Z_2 structure according to the binary theory of stopping for argon (upper graph) and helium (middle graph) ions in charge equilibrium, including projectile excitation, and bare protons (lower graph). Beam energy decreasing by factors of 10 from 10^3 MeV/u (top curve) to 10^{-3} MeV/u (bottom curve). From Sigmund & al. (2003).

The importance of projectile screening is further elucidated by a comparison of Fig. 6.5 with Fig. 5.5 on page 40, where Z_2 structure was found to *increase* with increasing Z_1 and increasing beam energy. This behavior is characteristic of the regime of small to negligible screening.

References

Arista, N. R. (2002). "Energy loss of heavy ions in solids: non-linear calculations for slow and swift ions," Nucl. Instrum. Methods B **195**, 91–105.

Betz, H. D. (1972). "Charge states and charge-changing cross sections of fast heavy ions penetrating through gaseous and solid media," Rev. Mod. Phys. **44**, 465–539.

Betz, H. D. and Grodzins, L. (1970). "Charge states and excitation of fast heavy ions passing through solids: a new model for the density effect," Phys. Rev. Lett. **25**, 211–214.

Bohr, N. (1940). "Scattering and stopping of fission fragments," Phys. Rev. **58**, 654–655.

Bohr, N. (1941). "Velocity-range relation for fission fragments," Phys. Rev. **59**, 270–275.

Bohr, N. and Lindhard, J. (1954). "Electron capture and loss by heavy ions penetrating through matter," Mat. Fys. Medd. Dan. Vid. Selsk. **28 no. 7**, 1–31.

Brandt, W. and Kitagawa, M. (1982). "Effective stopping-power charges of swift ions in condensed matter," Phys. Rev. B **25**, 5631–5637.

Brunelle, A., Della-Negra, S., Depauw, J., Jacquet, D., LeBeyec, Y. and Pautrat, M. (1997). "MeV cluster interactions with solids: explosion, charge states and secondary emission," Technical report IPNO-DRE-97-35, Institut de Physique Nucleaire, Orsay.

Fermi, E. and Amaldi, E. (1934). "Le orbite degli elementi," Mem. Accad. Italia **6**, 119–149.

Geissel, H., Weick, H., Scheidenberger, C., Bimbot, R. and Gardès, D. (2002). "Experimental studies of heavy-ion slowing down in matter," Nucl. Instrum. Methods B **195**, 3–54.

Glazov, L. G. (2002). "Frozen-charge stopping of ions in the Bethe regime," Nucl. Instrum. Methods B **195**, 118–132.

Grande, P. L. and Schiwietz, G. (2002). "The unitary convolution approximation for heavy ions," Nucl. Instrum. Methods B **195**, 55–63.

Ichihara, A., Shirai, T. and Eichler, J. (1993). "Cross sections for electron capture in relativistic atomic collisions," Atomic Data Nucl. Data Tables **55**, 63–79.

ICRU (2005). "Stopping of Heavy Ions," J. ICRU to appear.

Itoh, A., Tsuchida, H., Majima, T., Yogo, A. and Ogawa, A. (1999). "Equilibrium charge distributions of lithium ions emerging from a carbon foil," Nucl. Instrum. Methods B **159**, 22–27.

Kabachnik, N. M. (1993). "Screening and antiscreening in the semiclassical description of ionization in fast ion-atom collisions," J. Phys. B **26**, 3803–3814.

Kaneko, T. (1999). "Energy-loss of swift boron and carbon clusters in solids," Nucl. Instrum. Methods B **153**, 15–20.

Kim, Y. K. and Cheng, K. T. (1980). "Stopping power for partially stripped ions," Phys. Rev. A **22**, 61–67.

Knipp, J. and Teller, E. (1941). "On the Energy Loss of Heavy Ions," Phys. Rev. **59**, 659–669.

Knudson, A. R., Burghalter, P. G. and Nagel, D. J. (1974). "Vacancy configurations of argon projectile ions in solids," Phys. Rev. A **10**, 2118–2122.

Lassen, N. O. (1951a). "Total charges of fission fragments as functions of the pressure in the stopping gas," Mat. Fys. Medd. Dan. Vid. Selsk. **26 no. 12**, 1–19.

Lassen, N. O. (1951b). "The total charges of fission fragments in gaseous and solid stopping media," Mat. Fys. Medd. Dan. Vid. Selsk. **26 no. 5**, 1–28.

Lindhard, J. (1954). "On the properties of a gas of charged particles," Mat. Fys. Medd. Dan. Vid. Selsk. **28 no. 8**, 1–57.

Maynard, G., Chabot, M. and Gardès, D. (2000). "Density effect and charge dependent stopping theories for heavy ions in the intermediate velocity regime," Nucl. Instrum. Methods B **164-165**, 139–146.

Maynard, G., Katsonis, K. and Mabong, S. (1996). "Average atom model for swift heavy ions in dense matter," Nucl. Instrum. Methods B **107**, 51–55.

Maynard, G., Zwicknagel, G., Deutsch, C. and Katsonis, K. (2001). "Diffusion-transport cross section and stopping power of swift heavy ions - art. no. 052903," Phys. Rev. A **63**, 052903-1–14.

Nikolaev, V. S. and Dmitriev, I. S. (1968). "On the equilibrium charge distributions of heavy elemental ion beams," Phys. Lett. A **28**, 277–278.

Northcliffe, L. C. (1960). "Energy loss and effective charge of heavy ions in aluminum," Phys. Rev. **120**, 1744–1757.

Northcliffe, L. C. (1963). "Passage of heavy ions through matter," Ann. Rev. Nucl. Sci. **13**, 67–102.

Rozet, J. P., Stephan, C. and Vernhet, D. (1996). "ETACHA: a program for calculating charge states at GANIL energies," Nucl. Instrum. Methods B **107**, 67–70.

Sørensen, A. H. (2002). "Stopping of relativistic hydrogen- and heliumlike heavy ions," Nucl. Instrum. Methods B **195**, 106–117.

Schinner, A. and Sigmund, P. (2000). "Polarization effect in stopping of swift partially screened heavy ions: perturbative theory," Nucl. Instrum. Methods B **164-165**, 220–229.

Schiwietz, G. and Grande, P. L. (2001). "Improved charge-state formulas," Nucl. Instrum. Methods B **175-177**, 125–131.

Shima, K., Ishihara, T. and Mikumo, T. (1982). "Empirical formula for the average charge-state of heavy ions behind various foils," Nucl. Instrum. Methods **200**, 605–608.

Shima, K., Kuno, N., Yamanouchi, M. and Tawara, H. (1992). "Equilibrium charge fractions of ions of $Z = 4 - 92$ emerging from a carbon foil," Atom. Data Nucl. Data Tab. **51**, 173–241.

Sigmund, P. (1997). "Charge-dependent electronic stopping of swift nonrelativistic heavy ions," Phys. Rev. A **56**, 3781–3793.

Sigmund, P., Fettouhi, A. and Schinner, A. (2003). "Material dependence of electronic stopping," Nucl. Instrum. Methods B **209**, 19–25.

Sigmund, P. and Glazov, L. (1998). "Energy loss and charge exchange: statistics and atomistics," Nucl. Instrum. Methods B **136-38**, 47–54.

Sigmund, P. and Glazov, L. G. (2003). "Interplay of charge exchange and projectile excitation in the stopping of swift heavy ions," Europ. Phys. J. D **23**, 211–215.

Sigmund, P. and Schinner, A. (2000). "Binary stopping theory for swift heavy ions," Europ. Phys. J. D **12**, 425–434.

Sigmund, P. and Schinner, A. (2001a). "Effective charge and related/unrelated quantities in heavy-ion stopping," Nucl. Instrum. Methods B **174**, 535–540.

Sigmund, P. and Schinner, A. (2001b). "Nonperturbative theory of charge-dependent heavy-ion stopping," Phys. Scr. **T92**, 222–224.

Sigmund, P. and Schinner, A. (2002a). "Binary theory of electronic stopping," Nucl. Instrum. Methods B **195**, 64–90.

Sigmund, P. and Schinner, A. (2002b). "Binary theory of light-ion stopping," Nucl. Instrum. Methods B **193**, 49–55.

Weick, H., Sørensen, A. H., Geissel, H., Maier, M., Münzenberg, G., Nankov, N. and Scheidenberger, C. (2002). "Stopping power of partially ionized relativistic heavy ions," Nucl. Instrum. Methods B **193**, 1–7.

Yarlagadda, B. S., Robinson, J. E. and Brandt, W. (1978). "Effective-charge theory and the electronic stopping power of solids," Phys. Rev. B **17**, 3473–3483.

Ziegler, J. F., Biersack, J. P. and Littmark, U. (1985). "The stopping and range of ions in solids," J. F. Ziegler, ed., *The Stopping and Ranges of Ions in Matter*, volume 1 of *The Stopping and Ranges of Ions in Matter*, 1–319 (Pergamon, New York).

7 Aggregation Effects

7.1 Stopping in Compounds and Alloys, and Phase Effects

While stopping in elemental materials has been the dominating object of both experimental and theoretical studies, stopping in compounds and alloys is of prime interest in numerous applications.

A common reference standard is *Bragg's additivity rule*, which can be written in the form

$$-\frac{\mathrm{d}E}{\mathrm{d}\ell} = \sum_\nu n_\nu S_\nu, \tag{7.1}$$

where n_ν and S_ν are the number of atoms per volume and the stopping cross section, respectively, of the ν'th species of target atoms. This assumes that the stopping cross section of each species is unaffected by the state of aggregation. Hence the rule, when valid, pertains to both chemical and phase effects.

Deviations from additivity have received much attention. Systematic experimental studies for penetrating protons date back to the 1950s (Reynolds & al., 1953), and extensive reviews have been given by Thwaites (1983, 1984, 1987, 1992), by Ziegler and Manoyan (1988) and in ICRU (1993). The vast majority of available experimental data refers to He bombardment, especially by Powers & al. (1973) and numerous followup papers by Powers' group (Powers & al., 1984), and to Li bombardment by Pietsch & al. (1976) and followup papers by Neuwirth's group (Neuwirth and Both, 1985).

In particular, schemes were proposed for incorporating such deviations in a manner so that (7.1) can be maintained with a modified stopping cross section depending on the class of compound (ICRU, 1984, 1993).

In the high-velocity regime, as long as shell, Barkas-Andersen and screening corrections are unimportant, stopping forces are governed by the oscillator-strength spectrum. Here, (7.1) follows from (5.1) provided that optical oscillator strengths are additive,

$$f_j = \frac{\sum_\nu Z_\nu f_{j,\nu}}{\sum_\nu Z_\nu}, \tag{7.2}$$

where Z_ν and $f_{j,\nu}$ denote the atomic number and optical oscillator strength of the ν'th species. Since significant deviations from (7.2) can be expected mainly for valence and conduction electrons, deviations from Bragg's additivity rule are then governed by

- the variation of the oscillator-strength spectrum for valence electrons between the atomic and a compound state and
- the relative significance of those electrons in the total stopping force.

This situation is quite analogous to the Z_2 structure discussed in sections 5.7 and 8.7. In particular, one expects deviations from additivity to become increasingly pronounced with decreasing speed because of more rapid variation of the Bethe or Bohr logarithm with ω_j.

This trend is enhanced as shell corrections become important because of the closing of inner-shell excitation channels. However increasing significance of projectile screening with decreasing projectile speed will tend to wipe out this effect for heavier ions.

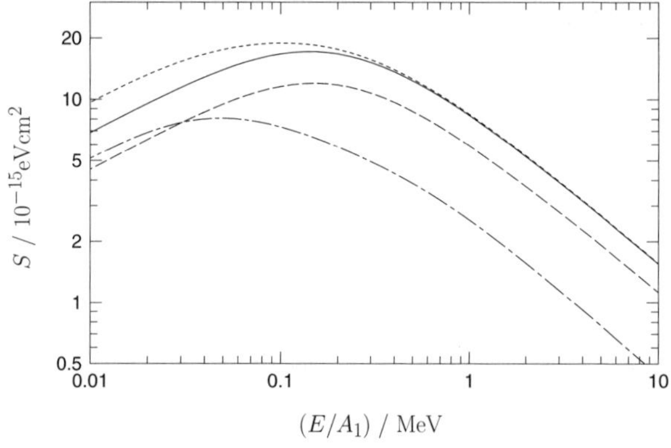

Fig. 7.1. Stopping cross section of lithium fluoride (solid line), lithium (dot-dashed line) and fluorine (dashed line) for antiprotons, calculated from binary theory. Also included is the sum of the elemental stopping cross sections (Bragg rule, dotted line). From Sharma & al. (2004).

Figures 7.1 and 7.2 illustrate these features on the case of LiF. This substance has been chosen because from a theoretical point of view it must be expected to show the most pronounced deviations from additivity:

- very large difference in binding energy of the outermost electron between metallic Li and LiF and
- a high fraction of outer electrons because of the low-Z_2 materials involved.

In Fig. 7.1 it is seen that for antiproton bombardment, in the complete absence of projectile screening, the predicted deviation from Bragg additivity

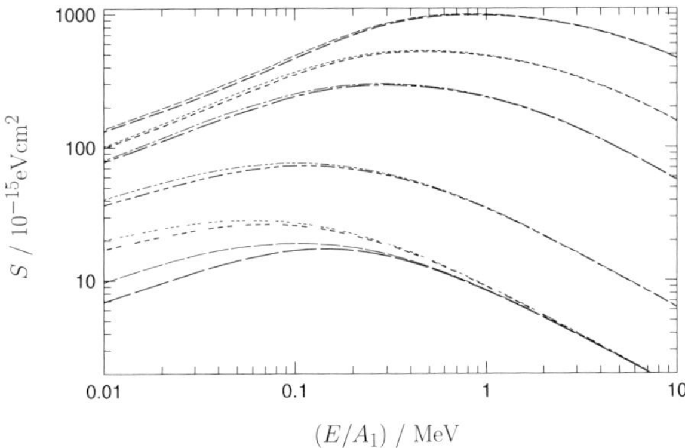

Fig. 7.2. Stopping cross section of lithium fluoride for argon, neon, carbon, helium, hydrogen ions and antiprotons (top to bottom, solid lines) calculated from binary theory. Also included are the sums of the elemental stopping cross sections (Bragg rule, broken lines). From Sharma & al. (2004)

reaches a factor of about 1.5 at low velocities. This is considerably more than what has been measured experimentally for any system. A direct experimental test would be of interest. Stopping measurements on metallic lithium have been shown to be possible (Eppacher & al., 1995).

Conversely, Fig. 7.2 shows that the effect decreases in importance and becomes insignificant from carbon upwards. In all cases it decreases rapidly above the stopping maximum.

On the basis of these estimates one may expect Bragg's rule to be adequate for estimating stopping forces on ions with $Z_1 \gtrsim 6$ above the Bohr velocity. However, application of Bragg's rule requires knowledge of the stopping cross sections of all constituent elements. In some cases where those are not known, stopping cross sections of compounds may still be estimated provided that adequate knowledge is available of the optical properties of the material.

7.2 Stopping of Molecules and Aggregates

As a first approximation the stopping force on a molecule, aggregate or cluster is given by the sum of stopping forces on its constituent atoms,

$$\left(\frac{\mathrm{d}E}{\mathrm{d}\ell}\right)_{\mathrm{mol}} = \sum \left(\frac{\mathrm{d}E}{\mathrm{d}\ell}\right)_{\mathrm{atom}} \tag{7.3}$$

at the same speed for both electronic and nuclear stopping[1]. Deviations from strict additivity have been found mainly in the electronic stopping of hydrogen molecules and clusters (Brandt & al., 1974; Ray & al., 1992). Similar effects for aggregates of heavier ions have been looked for experimentally and theoretically but were found to be only a few per cent (Baudin & al., 1994; Tomaschko & al., 1995; Ben-Hamu & al., 1997).

A theoretical analysis by Jensen and Sigmund (2000) showed that these findings are in agreement with stopping theory.

References

Baudin, K., Brunelle, A., Chabot, M., Della-Negra, S., Depauw, J., Gardès, D., kansson, P. H., LeBeyec, Y., Billebaud, A., Fallavier, M., Remillieux, J., Poizat, J. C. and Thomas, J. P. (1994). "Energy loss by MeV carbon clusters and fullerene ions in solids," Nucl. Instrum. Methods B **94**, 341–344.

Ben-Hamu, D., Baer, A., Feldman, H., Levin, J., Heber, O., Amitay, Z., Vager, Z. and Zajfman, D. (1997). "Energy loss of fast clusters through matter," Phys. Rev. A **56**, 4786–4794.

Brandt, W., Ratkowski, A. and Ritchie, R. H. (1974). "Energy loss of swift proton clusters in solids," Phys. Rev. Lett. **33**, 1325–1328.

Eppacher, C., Muiño, R. D., Semrad, D. and Arnau, A. (1995). "Stopping Power of Lithium for Hydrogen Projectiles," Nucl. Instrum. Methods B **96**, 639–642.

ICRU (1984). *Stopping powers for electrons and positrons*, volume 37 (ICRU Report, International Commission of Radiation Units and Measurements, Bethesda, Maryland).

ICRU (1993). *Stopping powers and ranges for protons and alpha particles*, volume 49 (ICRU Report, International Commission of Radiation Units and Measurements, Bethesda, Maryland).

Jensen, J. and Sigmund, P. (2000). "Electronic stopping of swift pastially stripped molecules and clusters," Phys. Rev. A **61**, 032903-1–14.

Neuwirth, W. and Both, G. (1985). "Aggregates of Atoms and their Stopping Cross Sections," Nucl. Instrum. Methods B **12**, 67–72.

Pietsch, W., Hauser, U. and Neuwirth, W. (1976). "Stopping powers from the inverted doppler shift attenuation method: Z-oscillations: Bragg's rule or chemical effects; solid and liquid state effects," Nucl. Instrum. Methods **132**, 79–87.

Powers, D., Lodhi, A. S., Lin, W. K. and Cox, H. L. (1973). "Molecular effects in the energy loss of alpha particles in gaseous media," Thin Sol. Films **19**, 205–215.

Powers, D., Olson, H. G. and Gowda, R. (1984). "Molecular Stopping Powers and Effect of Chemical Bonding in Gaseous Amine Compounds," J. Appl. Phys. **55**, 1274–1277.

---

[1] Although (7.3) is formally similar to (2.4) there is an essential difference. Equation (2.4) *defines* the energy loss of a composite particle via the energy loss experienced by its constituents, while (7.3) assumes that the energy loss of each constituent is equal to that experienced under *isolated* slowing-down.

Ray, E., Kirsch, R., Mikkelsen, H. H., Poizat, J. C. and Remillieux, J. (1992). "Slowing down of hydrogen clusters in thin foils," Nucl. Instrum. Methods B **69**, 133–141.

Reynolds, H. K., Dunbar, D. N. F., Wenzel, W. A. and Whaling, W. (1953). "The stopping cross section of gases for protons, 30-600 keV," Phys. Rev. **92**, 742.

Sharma, A., Fettouhi, A., Schinner, A. and Sigmund, P. (2004). "Electronic stopping of swift ions in compounds," Nucl. Instrum. Methods B in press.

Thwaites, D. I. (1983). "Bragg's rule of stopping power additivity: a compilation and summary of results," Radiat. Res. **95**, 495–518.

Thwaites, D. I. (1984). "Current status of physical state effects on stopping power," Nucl. Instrum. Methods B **12**, 84–89.

Thwaites, D. I. (1987). "Review of stopping powers in organic materials," Nucl. Instrum. Methods B **27**, 293–300.

Thwaites, D. I. (1992). "Departures from Bragg's rule of stopping power additivity for ions in dosimetric and related materials," Nucl. Instrum. Methods B **69**, 53–63.

Tomaschko, C., Brandl, D., Kügler, R., Schurr, M. and Voit, H. (1995). "Energy loss of MeV carbon cluster ions in matter," Nucl. Instrum. Methods B **103**, 407–411.

Ziegler, J. F. and Manoyan, J. M. (1988). "The stopping of ions in compounds," Nucl. Instrum. Methods B **35**, 215–228.

8 Low-Velocity Electronic Stopping

8.1 General Considerations

The regime of low-energy electronic stopping has been defined by

$$v \lesssim v_0 \quad \text{or} \quad E/A_1 \lesssim 25 \text{ keV} \tag{8.1}$$

in Fig. 4.2. The fact that the projectile speed is not large compared to even the lowest orbital speeds of the target electrons implies that 'sudden' Coulomb excitation underlying Bohr or Bethe theory ceases to be an effective energy-loss channel. Moreover, projectiles tend to be predominantly neutral in charge equilibrium, cf. (6.3).

While Bohr (1948) asserted that electronic-stopping cross sections would drop rapidly to zero for $v < v_0$, subsequent theoretical considerations led to a predicted friction-like behavior of the stopping force at low projectile speed. Such a behavior emerged from very different and mutually independent arguments.

8.2 Free Target Electrons

Fermi and Teller (1947), estimating slowing-down and capture of negative muons in matter, pointed out that the *rate* of energy loss, $-\mathrm{d}E/\mathrm{d}t$ of a slow heavy particle in a Fermi gas becomes proportional to its kinetic energy. This is equivalent to Stokes' law of a velocity-proportional stopping force.

The origin of this result is most easily identified by viewing the interaction in a reference frame moving along with the (heavy) projectile: When such a projectile is hit by a target electron, an amount of momentum is transferred that is proportional to the electron velocity. For an isotropic velocity distribution of target electrons these momentum transfers will cancel, but the small anisotropy induced by viewing the system from a slowly-moving reference frame causes a net momentum transfer proportional to and directed opposite to the projectile velocity.

A more quantitative version of this finding emerges from a comprehensive treatment of the stopping of a point charge in a Fermi gas by Lindhard (1954). In this formalism electronic properties of a material are described in terms of a frequency- and wavenumber-dependent dielectric function $\varepsilon(k, \omega)$. This allows for a self-consistent description of the response of an electron gas to a high degree of rigor, at least to the lowest order in the electric field induced by the projectile. The theory reproduces results of the Bethe (1930) theory at high projectile speed – with the plasma frequency replacing $I/\hbar$ – and a velocity-proportional stopping force at low speed dependent on density (Lindhard and Winther, 1964).

An important observation pointed out by Lindhard (1954) is the fact that it is not necessary to explicitly take into account the Pauli principle in the collision kinematics for a homogeneous electron gas, because for every scattering event that is forbidden by the Pauli principle there is another event with an equal but oppositely-directed momentum transfer at the same probability[1]. This is a key point in attempts to characterize the stopping behavior of inhomogeneous systems like atoms and molecules in terms of free-electron models and Thomas-Fermi-type arguments.

8.3 Bound Target Electrons

Independently and almost simultaneously, two qualitative approaches were developed that led to predictions of velocity-proportional stopping cross sections also for electrons bound in atoms. Firsov (1959) viewed the ion-atom collision system as a quasi-molecule with a flow of electrons between the collision partners. Energy loss is determined by the momentum needed to accelerate target electrons to the projectile velocity, and the number of electrons involved is determined by simple geometric considerations in combination with Thomas-Fermi-type arguments.

The goal of this work was to provide a qualitative explanation of the inelastic energy loss in low-energy ion-atom collisions. The mean inelastic energy loss w per collision at a given impact parameter p was given in the form

$$w(v, p) = 0.35 \frac{\hbar v}{a_0} \frac{(Z_1 + Z_2)^{5/3}}{\left[1 + 0.16(Z_1 + Z_2)^{1/3} r_{\min}/a_0\right]^5}$$

$$\text{for} \quad 1/4 \leq Z_1/Z_2 \leq 4, \quad (8.2)$$

where $r_{\min} = r_{\min}(v, p)$, is the closest distance of approach between the colliding nuclei.

Firsov was aware of the dominating role of electron promotion between quasimolecular orbitals in the process of electron excitation in slow ion-atom collisions, formulated in more explicit terms several years later by

---

[1]The Pauli principle does enter implicitly through the velocity distribution of the target electrons (Lindhard, 1954), and it cannot be neglected in straggling (Sigmund, 1982).

Fano and Lichten (1965). However, Firsov's theory did not aim at particle stopping. Application to this area dates back to Teplova & al. (1962) who integrated (8.2) over impact parameter – (re)identifying r_{min} with p so that

$$S = \int_0^\infty 2\pi p\,\mathrm{d}p\,w(p) = 2.3\,\pi a_0^2 (Z_1 + Z_2) m v_0 v. \tag{8.3}$$

A slightly modified version of the Firsov theory that, amongst other features, does not impose the above limitation on the range of Z_1/Z_2, is due to Kishinevskii (1962).

An alternative approach is due to Lindhard and Scharff (1961)[2]. This model determines momentum transfer in a quasi-molecule formed during collision on the basis of the Fermi-gas model mentioned above. This, in conjunction with a Thomas-Fermi description of the orbits of the scattering nuclei, leads to a stopping cross section of the form

$$S = \xi_e 8\pi a_0^2\,\frac{Z_1 Z_2}{\left(Z_1^{2/3} + Z_2^{2/3}\right)^{3/2}}\,m v_0 v, \tag{8.4}$$

where $\xi_e \simeq Z_1^{1/6}$ is an empirical parameter added to improve agreement with stopping measurements on fission fragments – the only pertinent data available at the time when the formula was established.

Figure 8.1 shows the ratio of the two expressions, indicating pronounced differences depending on atomic numbers involved.

Equation (8.4) has been successfully applied in the analysis of ion implantation profiles at keV and low-MeV energies, typically with a significant nuclear-stopping component.

8.4 Z_1 Structure: Modified-Firsov Models

Direct measurements of low-speed stopping on thin films and gas targets were performed by Teplova & al. (1962); Ormrod and Duckworth (1963); Ormrod & al. (1965); Fastrup & al. (1966); Ormrod (1968); Hvelplund and Fastrup (1968); Hvelplund (1971); Bierman & al. (1972); Hoffmann & al. (1976); Ward & al. (1979); Lennard & al. (1986) and Lennard and Geissel (1987). Pertinent experimental aspects in these and later measurements will be discussed in chapter 4 of ICRU (2005). While

---

[2]This work was part of a pioneering project on low-energy ion implantation. Publication of this series of papers was heavily delayed by external circumstances, but with the exception of the stopping formula (8.4), all major findings eventually appeared in full papers (Lindhard & al., 1963a,b, 1968). A major reason for the missing published derivation of the stopping formula was the experimental discovery of Z_1 structure in low-energy stopping. Arguments in support of (8.4) were published by Sigmund (1975) and Tilinin (1995).

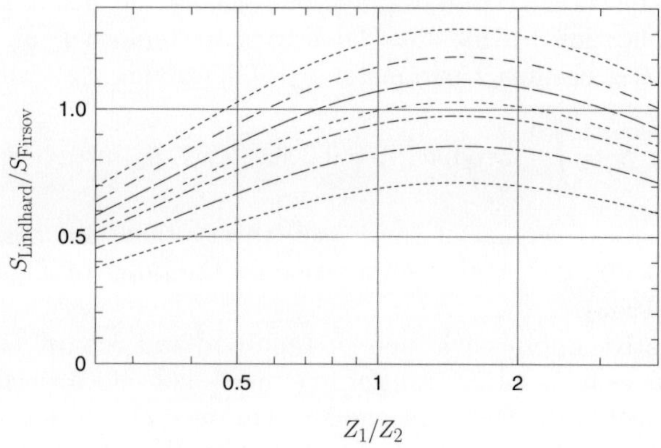

Fig. 8.1. Ratio between low-speed electronic stopping cross sections due to Lindhard and Scharff (1961) and Firsov (1959) for $Z_2 = 79, 47, 26, 18, 13, 6, 2$ (top to bottom). ξ_e has been set to $Z_1^{1/6}$. The interval covered for Z_1/Z_2 is limited by the range of validity of (8.3).

these measurements confirmed the general behavior predicted by (8.3) and (8.4), a distinct nonmonotonic behavior, 'Z_1 oscillations' or 'Z_1 structure' was found which had not been predicted theoretically.

Figure 8.2 shows data for a wide range of ions on amorphous carbon. Similar, less comprehensive data exist on aluminium, silicon, nickel, silver, gold, neon, argon, and air. Maxima and minima, where identifiable, lie at approximately the same values of Z_1 for these targets, and oscillation amplitudes range up to 15-20 %.

Much more pronounced Z_1 structure was found in the stopping of well-channeled ions in single crystals of tungsten (Eriksson & al., 1967), silicon (Eisen, 1968) and gold (Bøttiger and Bason, 1969), where a narrower range of impact parameters is sampled.

Initial attempts to explain these oscillations were based on the models of Firsov (El-Hoshi and Gibbons, 1968; Winterbon, 1968; Cheshire & al., 1968) or Lindhard (Bhalla and Bradford, 1968) with modified electron densities, electron fluxes, and projectile charge states. Most of these and numerous subsequent approaches along similar lines (Bhalla & al., 1970; Cheshire and Poate, 1970; Kessel'man, 1971b,a; Bierman & al., 1972; Baklitsky & al., 1973; Komarov and Kumakhov, 1973) were reasonably successful in the prediction of the positions of maxima and minima, and some of them also matched the amplitudes, although usually with the help of adjustable parameters.

None of these approaches has been utilized to systematically produce *theoretical predictions*, and none of them has been capable of explaining observed deviations from strictly velocity-proportional stopping. The latter feature is

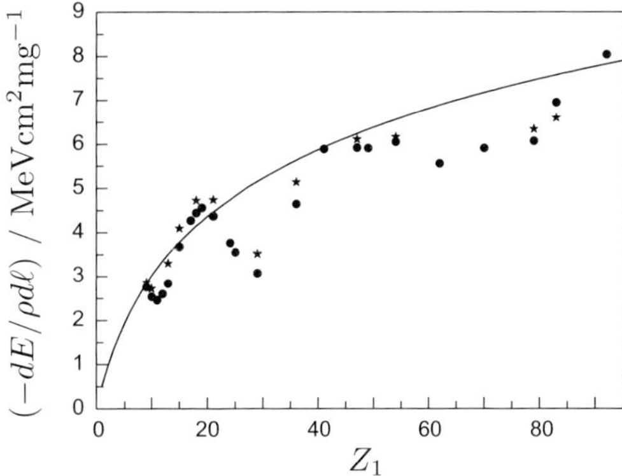

Fig. 8.2. Measured Z_1 oscillations in carbon at 0.8 v_0. Solid line: Equation (8.4). Data from Lennard & al. (1986). Filled circles: thickness 9 µg/cm^2; stars: thickness 29 µg/cm^2.

usually summarized in an empirical power law, $S \propto v^\mathsf{p}$, where p was found to oscillate around 0.5 as a function of Z_1 at considerable amplitudes, e.g. between 0.3 and 0.9 for stopping along the $\langle 110 \rangle$ channel in silicon (Eisen, 1968). Numerical results were found to be sensitive to detailed input, a feature that has given rise to some discussion (Cruz & al., 1979; Komarov, 1979).

8.5 Z_1 Structure: Lindhard-Finnemann Model

An alternative approach proposed by Lindhard[3] explains the Z_1 structure as a quantum effect. Stopping is described in terms of the transport cross section (5.10), which is governed by the phase shifts for scattering of low-energy electrons ($v < v_0$) on the screened Coulomb potential of a neutral or nearly-neutral projectile atom. In this velocity range, low values of ℓ dominate the scattering amplitude. This leads to maxima and minima in the stopping cross section near the point where δ_0 approaches uneven or even multiples of $\pi/2$. In addition to the scattering potential, phase shifts also depend on electron speed. This is known as the Ramsauer-Townsend effect in the scattering of free electrons on gas atoms.

The evaluation by Finnemann (1968) of WKB (Wentzel-Kramers-Brillouin) phase shifts on a Lenz-Jensen potential predicted stopping maxima at Z_1=6, 20, 41 and 72 and minima at $Z_1 = 12$, 29 and 55 for $v = v_0$. At $v = v_0/2$, maxima were found at $Z_1 = 7$, 18, 37 and 68 and minima at $Z_1 =$

---

[3]Lindhard's work dates back to 1968 but was never published. A written record is available in an M.Sc. thesis (Finnemann, 1968). The first published record is by Briggs and Pathak (1974).

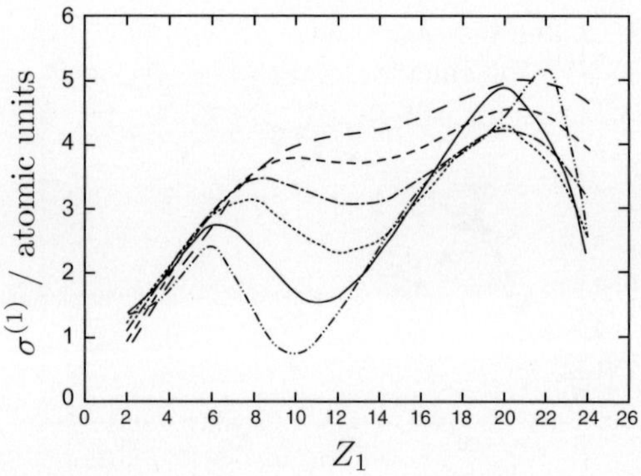

Fig. 8.3. Predicted velocity dependence of transport cross section $\sigma^{(1)}$, (5.10) in a Fermi gas according to Pathak (1980). $v/v_0 = 0.75$, 1.0, 1.25, 1.5, 1.75, 2.0 (bottom to top).

3, 11, 26, 51 and 89, in fair agreement with experimental data shown in Fig. 8.2.

Briggs and Pathak (1974) found similar results on the basis of a Molière potential and numerically-evaluated phase shifts. Further calculations on the basis of Hartree-Fock atomic densities produced additional structure which was asserted to be sensitive to projectile speed and likely to be wiped out in a more realistic treatment of electron scattering in a solid medium.

The scheme was analysed in a considerable number of followup studies (Briggs and Pathak, 1974; Pathak, 1974a; Ali and Gallaher, 1974; Pathak, 1980; Kumar and Pathak, 1993) with applications mostly to channeling data on Si, W and Au. Good agreement has been reached with experimental results in particular by Kumar and Pathak (1993). The velocity dependence of Z_1 structure was studied in this model by Pathak (1980). Figure 8.3 shows the transport cross section for a free-electron gas as a function of Z_1 for a speed ratio v/v_0 increasing from 0.75 to 2.0. It is seen that the oscillation amplitude decreases rapidly. Moreover, the position of maxima and minima moves toward higher Z_1 with increasing speed, in accordance with the prediction of Finnemann (1968) mentioned above.

8.6 Self-Consistent Nonlinear Models

A key parameter in the calculations discussed above is the screening potential accompanying the projectile, which was determined from a more or less sophisticated atomic-charge distribution in all cases. At low projectile speeds this potential is modified by the presence of a sea of target electrons. A self-consistent theoretical description requires taking into account the relaxation of the electron system due to the presence of the penetrating ion. Scattering phase shifts then need to be calculated with the resulting self-consistent potential as input. One test for self-consistency is the Friedel sum rule which reads

$$\frac{2}{\pi} \sum_\ell (2\ell + 1)\delta_\ell(E_F) = Z_1 \tag{8.5}$$

for a static impurity atom Z_1 embedded into a Fermi gas with Fermi energy E_F.

The dielectric description of Lindhard (1954) satisfies the requirement of self-consistency up to the first order in the electric field. The density-functional theory of Hohenberg and Kohn (1964) and Kohn and Sham (1965) goes significantly beyond this approximation. The use of this scheme in the evaluation of low-speed stopping in a homogeneous electron gas was initiated by Echenique & al. (1981) for protons and helium ions. Z_1-dependent stopping was studied by Echenique & al. (1986). A considerable number of followup studies (Ashley & al., 1986; Arnau & al., 1988; Arnau and Echenique, 1989; Echenique & al., 1991; Peñalba & al., 1992) was devoted to the analysis of Z_1 structure. Estimates by Ashley & al. (1986) and Echenique & al. (1991) — who modelled the undisturbed target as a homogeneous Fermi gas — reproduced observed trends for amorphous carbon and channeling in silicon and gold. Estimates by Peñalba & al. (1992), based on more realistic electron distributions in the target, were in good agreement with measurements for $\langle 110 \rangle$ Si.

A much simpler approach was presented by Calera-Rubio & al. (1994), where a Yukawa-type screening function was adopted with a screening radius determined recursively by the Friedel sum rule via calculated scattering phase shifts. This approach was further developed by establishing a generalized Friedel sum rule (Lifschitz and Arista, 1998). The resulting theory of heavy-ion stopping that summarizes much of the development described in this section (Arista, 2002) will be described in Sect. 9.3.

8.7 Z_2 Structure

Unlike Z_1 structure, Z_2 structure is present at all velocities, although most pronouncedly so in the low-v regime as discussed in Sect. 5.7. It is of interest, therefore, to explore what genuine low-speed stopping theory can contribute to this topic.

Although all theoretical models discussed above allow predictions on Z_2 structure, systematic studies are scarce. Attempts by Pathak (1974b) and Ali and Gallaher (1974) addressed channeled ions, directed at range measurements by Whitton (1974). Latta and Scanlon (1976) made predictions on the basis of a modified-Firsov model, aiming at experimental results by Broude & al. (1972). Pietsch & al. (1976), in analysing their measurements of low-speed stopping of Li, applied a modification to the Lindhard-Scharff formula (8.4). This model and two modified-Firsov models were employed in the analysis of data on low-speed stopping by Land $\&$ al. (1977), with moderate success. Systematic studies on low-speed Z_2 structure on the basis of the quantum model would be desirable. While predictions are possible on the basis of binary theory, there are clear limitations in view of the fact that Z_1 structure is unaccounted for.

8.8 Conclusions

A number of conclusions may be drawn from this fairly large body of theoretical studies:

- Unlike the modified-Firsov models, the quantal model leads to deviations from velocity-proportional stopping. These deviations have not been studied quantitatively, and it is not known whether calculated velocity dependences match those measured.
- Models described above have been successfully utilized in estimates of Z_1 structure in energy losses of channeled ions. Particularly good agreement with measurements has been obtained with the quantal model both in the linear and the nonlinear version. This may be due to the fact that pertinent measurements concern 'best-channeled' ions which move in rather well-defined trajectories in regions with a fairly constant electron density described well by a homogeneous-electron-gas model.
- For *random* slowing-down, really good agreement with existing measurements on Z_1 structure is not found for any of these models, at least as far as amplitudes are concerned. 'Best' estimates were presented by Komarov and Kumakhov (1973) and Calera-Rubio $\&$ al. (1994) for carbon, and by Calera-Rubio $\&$ al. (1994) for aluminium. However, unlike Komarov and Kumakhov (1973), Calera-Rubio $\&$ al. (1994) as well as Echenique $\&$ al. (1986) predicted oscillation amplitudes that are *larger* than those measured. This leaves open the possibility that oscillation amplitudes are underestimated in the analysis of experimental data, e.g., due to uncertainties in the nuclear-stopping correction.
- In addition to a pronounced dependence on projectile speed of predicted maximum and minimum positions, predicted Z_1 structure also depends on the target material according to Calera-Rubio $\&$ al. (1994), although the difference between the two cases studied (C and Al) is small.

References

Ali, S. P. and Gallaher, D. F. (1974). "Electronic stopping power of channelled ions in a model incorporating Pauli principle," J. Phys. C **7**, 2434–2446.

Arista, N. R. (2002). "Energy loss of heavy ions in solids: non-linear calculations for slow and swift ions," Nucl. Instrum. Methods B **195**, 91–105.

Arnau, A. and Echenique, P. M. (1989). "Stopping power of an electron gas for partially stripped ions," Nucl. Instrum. Methods B **42**, 165–170.

Arnau, A., Echenique, P. M. and Ritchie, R. H. (1988). "Stopping power of slow ions in metals and insulators," Nucl. Instrum. Methods B **33**, 138–141.

Ashley, J. C., Ritchie, R. H., Echenique, P. M. and Nieminen, R. M. (1986). "Nonlinear calculations of the energy loss of slow ions in an electron gas," Nucl. Instrum. Methods B **15**, 11–13.

Baklitsky, B. E., Parilis, E. S. and Ferleger, V. K. (1973). "Dependence of inelastic energy loss on the atomic number of the ions," Radiat. Eff. **19**, 155–160.

Bethe, H. (1930). "Zur Theorie des Durchgangs schneller Korpuskularstrahlen durch Materie," Ann. Physik **5**, 324–400.

Bhalla, C. P. and Bradford, J. N. (1968). "Oscillating behavior of electron stopping power," Phys. Lett. A **27**, 318–319.

Bhalla, C. P., Bradford, J. N. and Reese, G. (1970). "Critical examination of modified Firsov theory of inelastic energy loss in atomic collisions," D. W. Palmer, M. W. Thompson and P. D. Townsend, eds., *Atomic collisions in solids*, 361 (North Hollan, Amsterdam).

Bierman, D. J., Turkenburg, W. C. and Bhalla, C. P. (1972). "Inelastic energy losses in small-angle scattering of energetic particles," Physica **60**, 357–374.

Bohr, N. (1948). "The penetration of atomic particles through matter," Mat. Fys. Medd. Dan. Vid. Selsk. **18 no. 8**, 1–144.

Bøttiger, J. and Bason, F. (1969). "Energy loss of heavy ions along low-index directions in gold single crystals," Radiat. Eff. 10–31.

Briggs, J. S. and Pathak, A. P. (1974). "The stopping power of solids for low-velocity channelled heavy ions," J. Phys. C **7**, 1929–1936.

Broude, C., Engelstein, P., Popp, M. and Tandon, P. N. (1972). "Dependence of the Doppler shift lifetime method on slowing environment," Phys. Lett. B **39**, 185–187.

Calera-Rubio, J., Gras-Marti, A. and Arista, N. R. (1994). "Stopping power of low-velocity ions in solids - inhomogeneous electron-gas model," Nucl. Instrum. Methods B **93**, 137–141.

Cheshire, I. M., Dearnaley, G. and Poate, J. M. (1968). "The Z_1-dependence of electronic stopping," Phys. Lett. A **27**, 304–305.

Cheshire, I. M. and Poate, J. M. (1970). "Shell effects in low-energy atomic collisions," D. W. Palmer, M. W. Thompson and P. D. Townsend, eds., *Atomic collisions in solids*, 351–360 (North Hollan, Amsterdam).

Cruz, S. A., Vargas, C. and Brice, D. K. (1979). "Critical analysis of the modified Firsov model. Sensitivity to the choice of atomic wavefunctions," Radiat. Eff. Lett. **43**, 79–84.

Echenique, P. M., Arnau, A., Peñalba, M. and Nagy, I. (1991). "Stopping power of low velocity ions in solids," Nucl. Instrum. Methods B **56-57**, 345–347.

Echenique, P. M., Nieminen, R. M., Ashley, J. C. and Ritchie, R. H. (1986). "Nonlinear stopping power of an electron gas for slow ions," Phys. Rev. A **33**, 897–904.

Echenique, P. M., Nieminen, R. M. and Ritchie, R. H. (1981). "Density functional calculation of stopping power of an electron gas for slow ions," Sol. St. Comm. **37**, 779–781.

Eisen, F. H. (1968). "Channeling of medium-mass ions through silicon," Can. J. Phys. **46**, 561–572.

El-Hoshi, A. H. and Gibbons, J. F. (1968). "Periodic dependence of the electronic stopping cross section for energetic heavy ions in solids," Phys. Rev. **173**, 454–460.

Eriksson, L., Davies, J. A. and Jespersgaard, P. (1967). "Range measurements in oriented tungsten single crystals (0.1 – 1.0 MeV). I. Electronic and nuclear stopping powers," Phys. Rev. **161**, 219–234.

Fano, U. and Lichten, W. (1965). "Interpretation of Ar^+ – Ar collisions at 50 keV," Phys. Rev. Lett. **14**, 627–629.

Fastrup, B., Hvelplund, P. and Sautter, C. A. (1966). "Stopping cross section in carbon of 0.1-1.0 MeV atoms with $6 < Z_1 < 20$," Mat. Fys. Medd. Dan. Vid. Selsk. **35 no. 10**, 1–28.

Fermi, E. and Teller, E. (1947). "The capture of negative mesotrons in matter," Phys. Rev. **72**, 399–408.

Finnemann, J. (1968). *En redegørelse for resultaterne af beregninger over spredning af elektroner med lav energi på afskærmede Coulombfelter*, master thesis, Aarhus University.

Firsov, O. B. (1959). "A qualitative interpretation of the mean electron excitation energy in atomic collsions," Zh. Eksp. Teor. Fiz. **36**, 1517–1523, [English translation: Sov. Phys. JETP **9**, 1076-1080 (1959)].

Hoffmann, I., Jäger, E. and Müller-Jahreis, U. (1976). "Z_1-dependence of electronic energy straggling of light ions," Radiat. Eff. **31**, 57.

Hohenberg, P. and Kohn, W. (1964). "Inhomogeneous electron gas," Phys. Rev. **136**, B864–B871.

Hvelplund, P. (1971). "Energy loss and straggling of 100-500 keV atoms with $2 \leq Z_1 \leq 12$ in various gases," Mat. Fys. Medd. Dan. Vid. Selsk. **38 no. 4**, 1–25.

Hvelplund, P. and Fastrup, B. (1968). "Stopping cross sections in carbon of 0.2 - 1.5 MeV atoms with $21 \leq Z_1 \leq 39$," Phys. Rev. **165**, 408–414.

ICRU (2005). "Stopping of Heavy Ions," J. ICRU to appear.

Kessel'man, V. S. (1971a). "Oscillatory dependence on the atomic number of the projectile atom for the stopping power due to inelastic collisions," Zh. Tekh. Fiz. **42**, 1161.

Kessel'man, V. S. (1971b). "Stopping power of crystals for inelastic collisions with oscillatory dependence on ion charge," Zh. Tekh. Fiz. **42**, 1161, [English translation: Sov. Phys. Techn. Phys. **16**, 1346 (1972)].

Kishinevskii, L. M. (1962). "Cross sections for inelastic atomic collisions," Izv. Akad. NAUK SSSR **26**, 1410, [English translation: Bull. Acad. Sci. USSR Phys. Ser. **20**, 1433-1438 (1963)].

Kohn, W. and Sham, L. J. (1965). "Self-consistent equations including exchange and correlation effects," Phys. Rev. **140**, A1133–A1138.

Komarov, F. F. (1979). "Comment on "Critical analysis of the modified Firsov model. Sensitivity to the choice of atomic wave functions" by S. A. Cruz, C. Vargas and D. K. Brice," Radiat. Eff. Lett. **43**, 139.

Komarov, F. F. and Kumakhov, M. A. (1973). "Electronic energy loss of ions in the modified Firsov theory," phys. stat. sol. B **58**, 389–400.

Kumar, V. H. and Pathak, A. P. (1993). "Z_1-oscillations in the stopping powers of silicon and tungsten for low-velocity channelled heavy-ions," J. Phys.-Cond. Matter **5**, 3163–3168.

Land, D. J., Brennan, J. G., Simons, D. G. and Brown, M. D. (1977). "Comparison of theoretical models for the electronic stopping power of low-velocity heavy ions," Phys. Rev. A **16**, 492–499.

Latta, B. M. and Scanlon, P. J. (1976). "Average electronic energy loss in the modified Firsov theory," phys. stat. sol. B **74**, 711–719.

Lennard, W. N. and Geissel, H. (1987). "Energy loss and energy loss straggling for heavy ions," Nucl. Instrum. Methods B **27**, 338–343.

Lennard, W. N., Geissel, H., Jackson, D. and Phillips, D. (1986). "Electronic stopping values for low velocity ions ($9 \leq Z_1 \leq 92$) in carbon targets," Nucl. Instrum. Methods B **13**, 127–132.

Lifschitz, A. F. and Arista, N. (1998). "Velocity-dependent screening in metals," Phys. Rev. A **57**, 200–207.

Lindhard, J. (1954). "On the properties of a gas of charged particles," Mat. Fys. Medd. Dan. Vid. Selsk. **28 no. 8**, 1–57.

Lindhard, J., Nielsen, V. and Scharff, M. (1968). "Approximation method in classical scattering by screened coulomb fields," Mat. Fys. Medd. Dan. Vid. Selsk. **36 no. 10**, 1–32.

Lindhard, J., Nielsen, V., Scharff, M. and Thomsen, P. V. (1963a). "Integral equations governing radiation effects," Mat. Fys. Medd. Dan. Vid. Selsk. **33 no. 10**, 1.

Lindhard, J. and Scharff, M. (1961). "Energy dissipation by ions in the keV region," Phys. Rev. **124**, 128–130.

Lindhard, J., Scharff, M. and Schiøtt, H. E. (1963b). "Range concepts and heavy ion ranges," Mat. Fys. Medd. Dan. Vid. Selsk. **33 no. 14**, 1.

Lindhard, J. and Winther, A. (1964). "Stopping power of electron gas and equipartition rule," Mat. Fys. Medd. Dan. Vid. Selsk. **34 no. 4**, 1–22.

Ormrod, J. H. (1968). "Low-energy electronic stopping cross sections in nitrogen and argon," Can. J. Phys. **46**, 497–502.

Ormrod, J. H. and Duckworth, H. E. (1963). "Stopping cross sections in carbon for low-energy atoms with $Z \leq 12$," Can. J. Phys. **41**, 1424–1442.

Ormrod, J. H., MacDonald, J. R. and Duckworth, H. E. (1965). "Some low-energy atomic stopping cross sections," Can. J. Phys. **43**, 275–284.

Pathak, A. P. (1974a). "The Z_1 oscillations in stopping power of metals for low velocity channeled heavy-ions," J. Phys. F **4**, 1883–1888.

Pathak, A. P. (1974b). "Z_2 dependence of electronic stopping power of low-velocity channeled heavy-ions," J. Phys. C **7**, 3239–3244.

Pathak, A. P. (1980). "Systematic study of channeling stopping-power oscillations for low-velocity heavy ions," Phys. Rev. B **22**, 96–98.

Peñalba, M., Arnau, A. and Echenique, P. M. (1992). "Z_1 oscillations in slow channeled ion stopping power," Nucl. Instrum. Methods B **67**, 66–68.

Pietsch, W., Hauser, U. and Neuwirth, W. (1976). "Stopping powers from the inverted doppler shift attenuation method: Z-oscillations: Bragg's rule or chemical effects; solid and liquid state effects," Nucl. Instrum. Methods **132**, 79–87.

Sigmund, P. (1975). "Energy loss of charged particles in solids," C. H. S. Dupuy, ed., *Radiation damage processes in materials*, NATO Advanced Study Institutes Series, 3–117 (Noordhoff, Leyden).

Sigmund, P. (1982). "Kinetic theory of particle stopping in a medium with internal motion," Phys. Rev. A **26**, 2497–2517.

Teplova, Y. A., Nikolaev, V. S., Dimitriev, I. S. and Fateeva, L. N. (1962). "Slowing down of multicharged ions in solids and gases," Zh. Eksp. Teor. Fiz. **42**, 44–60, [English translation: Sov. Phys. JETP **15**, 31-41 (1962)].

Tilinin, I. S. (1995). "Quasiclassical expression for inelastic energy losses in atomic particle collisions below the Bohr velocity," Phys. Rev. A **51**, 3058–3065.

Ward, D., Andrews, H. R., Mitchell, I. V., Lennard, W. N., Walker, R. B. and Rud, N. (1979). "Systematics for the Z_1-oscillation in stopping powers of varioussolid materials," Can. J. Phys. **57**, 645–656.

Whitton, J. L. (1974). "The dependence of electronic stopping cross-section of K-42 an different target materials," Can. J. Phys. **52**, 12–16.

Winterbon, K. B. (1968). "Z_1 oscillations in stopping of atomic particles," Can. J. Phys. **46**, 2429–2433.

9 Survey of Current Theoretical Schemes

With a view to the growing interest in heavy-ion stopping, theoretical schemes have been developed over the past few years aiming at calculating stopping forces more or less from first principles. This section presents four currently available schemes, all of which are still under development.

9.1 Unitary-Convolution Approximation

The guiding principle behind the unitary-convolution approximation by Grande and Schiwietz (1998) is to provide an impact-parameter-dependent version of the Bloch theory. The energy loss of bare ions is determined by interpolation between close and distant interactions, where the latter are described by the standard scheme (dipole approximation) while close collisions are characterized by free-Coulomb interaction with an effective minimum impact parameter dependent on the Bohr parameter $\kappa = 2Z_1 v_0/v$. Interpolation is performed such as to reproduce the Bloch function after integration.

The scheme was extended by Azevedo & al. (2000) to screened ions by introduction of a screened potential, which in turn was found from Hartree-Fock type atomic and ionic projectile wave functions.

In its present stage (Grande and Schiwietz, 2002) the theory is geared to cover the transition from the Born to the classical regime. In addition to projectile screening, also projectile excitation and ionization are allowed for, although the contribution from projectile ionization is likely to be overestimated, cf. the remarks in Sect. 6.7 on page 53. Shell and Barkas-Andersen as well as relativistic corrections are omitted. For oxygen on aluminium this would suggest the theory to be valid within the range 0.5 MeV $\ll E/A_1 \ll 1$ GeV. This is confirmed in Fig. 9.1. The excellent agreement with experiment around 0.1 MeV may be assumed to be accidental (Grande and Schiwietz, 2002).

The model provides impact-parameter-dependent energy losses that must be expected to have comparable accuracy as the corresponding stopping cross sections. The potential of the theory has been explored mainly for He and Li bombardment (Azevedo & al., 2001; Grande & al., 2002), although the major issue of those studies, the Barkas-Andersen effect, was treated as input rather than output.

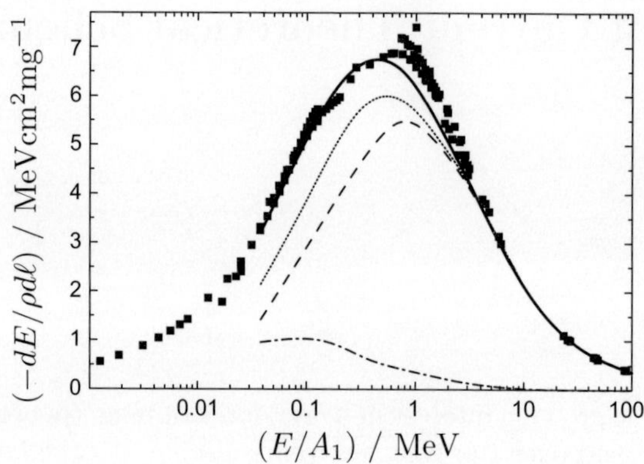

Fig. 9.1. Equilibrium stopping force for O-Al predicted from the unitary-convolution approximation, compared with experimental values from Fig. 4.1. Dashed line: Target excitation assuming mean ion charge. Dotted line: Target excitation averaged over ionic charge spectrum. Dot-dashed line: Projectile excitation. Solid line: Total stopping force averaged over ionic charge spectrum. From Grande and Schiwietz (2002).

Also charge-dependent ('frozen-charge') stopping cross sections were evaluated and compare favorably with experiment (Grande and Schiwietz, 2002).

The theory has been implemented in the program CasP, a somewhat reduced version of which – not allowing for projectile excitation/ionization – is available on the internet (Grande and Schiwietz, 2001).

9.2 Binary Theory

The physical model underlying binary stopping theory by Sigmund and Schinner (2000) is very close to that of Bohr (1913), but application of perturbation theory and a formal distinction between close and distant interactions has been avoided. This has been achieved by treating the effect of electron binding as screening of the interaction. In this way, a complex many-body problem (involving as a minimum the projectile and target nuclei and a target electron) has been mapped on a binary scattering problem involving the projectile and a target electron.

An estimate of the Barkas-Andersen effect is inherent in the picture, and shell corrections have been incorporated separately (Sigmund and Schinner, 2001b) by means of the transformation (5.7) that is rigorous for binary collisions. An essential ingredient is the inverse-Bloch correction mentioned

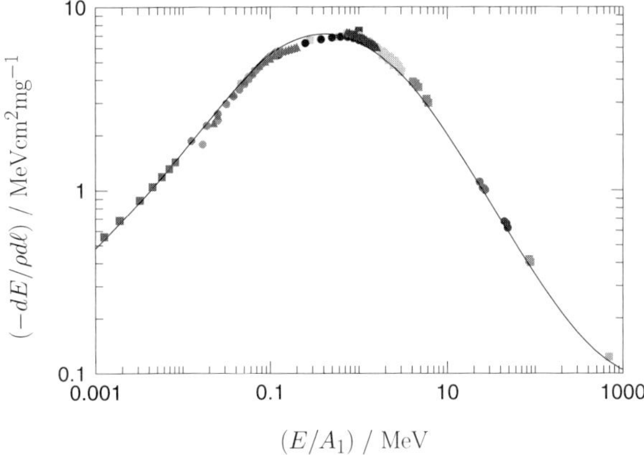

Fig. 9.2. Stopping of oxygen in aluminium: Comparison
of experimental data with prediction from binary theory.
From Sigmund and Schinner (2002a).

in Sect. 5.3, which extends the range of validity into the Born regime.
The theoretical scheme has been implemented in the program PASS which
includes the standard relativistic correction in (5.8), as well as the correction by Lindhard and Sørensen (1996), and allows for projectile ex-
citation/ionization. Details of the binary theory have been specified by
Sigmund and Schinner (2002a), and an up-to-date account will be given in
Chap. 6 of ICRU (2005).

Since optical oscillator-strength spectra form the main input into the theory, the accuracy of its predictions hinges on the quality of available optical
properties (refraction indices and attenuation coefficients) over an energy
range from about 1 eV to 10-100 keV, dependent on atomic number.

Extensive tests on the sensitivity to various types of input and comparisons with experimental data were presented by Sigmund and Schinner
(2002a). Figure 9.2 shows the case of O-Al with the experimental data also
shown in Fig. 9.1. Also charge-dependent stopping cross sections were evaluated and compare favorably with experiment (Sigmund and Schinner, 2001b).

Evidence presented by Sigmund and Schinner (2000) indicated that the
estimate of the Barkas-Andersen effect inherent in the binary theory can be
quite accurate. This conclusion has been strengthened by an evaluation of
antiproton stopping forces where excellent agreement has been achieved with
numerous experimental results (Sigmund and Schinner, 2001a, 2002b). An
example is shown in Fig. 5.4.

Although the theory is geared toward beam velocities exceeding v_0, Fig.
9.2 shows that good agreement with experimental results can be achieved
at lower velocities. However, the theoretical scheme does not incorporate a
model of the type described in Sects. 8.4 – 8.6 to predict Z_1 structure.

9.3 Nonlinear Electron-Gas Model

Stopping models based on the transport-cross-section approach in conjunction with scattering phase shifts, (5.10), have been common in low-speed stopping as discussed in sect. 8. The approach has also been highly successful in relativistic heavy-ion stopping, cf. sect. 5.3.

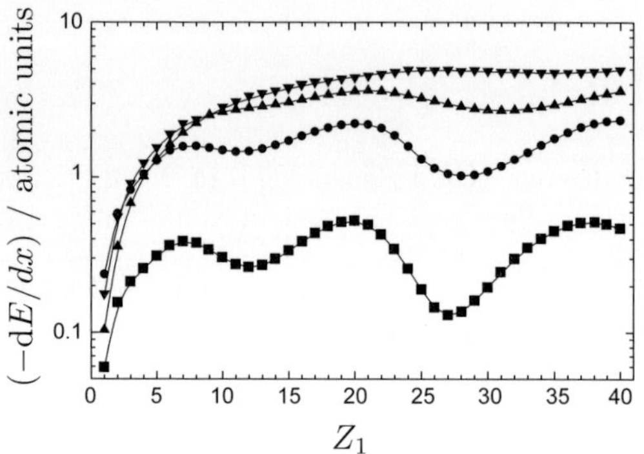

Fig. 9.3. Z_1 oscillation in stopping power as a function of projectile speed according to Arista (2002), calculated for homogeneous electron gas with a Wigner-Seitz radius $r_s = a_0(3/4\pi n_e)^{1/3} = 1.6$ representing carbon (n_e = electron density) for $v/v_0 = 3, 2, 1, 0.2$ (top to bottom). From Arista (2002)

An attempt to extend the range of validity of a low-speed scheme toward intermediate velocities was initiated by Lifschitz and Arista (1998). The scheme operates with a Fermi gas as a target, and its main ingredient is a generalized Friedel sum rule that takes into account the motion of the projectile through the medium. Apart from this, the ingredients of the theory – when applied to the stopping of point charges – are essentially the same as in the low-speed approach by Calera-Rubio & al. (1994). The scheme was applied to estimate antiproton stopping (Arista and Lifschitz, 1999, 2002). Results shown in Fig. 5.4 confirm that the theory incorporates a reasonable estimate of the Barkas-Andersen effect.

An extension to heavier ions was presented by Arista (2002). Particular attention was given to projectile screening, for which a variety of screening functions was explored. Being based on a Fermi gas, the scheme incorporates shell corrections from the outset while projectile excitation has not been included.

Figure 9.3 shows a general behavior very similar to Fig. 8.3 and confirms the fading away of predicted Z_1 oscillations with increasing projectile speed.

Electron-gas models are powerful in the description of low-speed stopping in normal metals, where stopping is predominantly due to quasi-free electrons. With increasing projectile speed, bound target electrons contribute to stopping. This requires separate attention. The problem gets accentuated for insulators. The problem was solved for antiproton stopping by application of some simplified linear model to inner shells (Arista and Lifschitz, 1999) or by application of the local-density approximation (Arista and Lifschitz, 2002).

9.4 CKT and Related Theories

Convergent kinetic theory (CKT) is a label for a number of extensions to the Bethe theory that allow stopping calculations for partially-ionized ions in partially-ionized targets. The theory is geared toward stopping in plasmas but reference is extensively made to cold matter, in particular in the work of Maynard & al. (1996, 1998b,a, 2000, 2001a,b, 2002b) summarized by Maynard & al. (2002a).

Due to the emphasis on applications to fusion plasmas, particular attention has been given to projectile processes, especially charge states and the correlation between charge exchange and energy loss. A high degree of symmetry has been aimed at with regard to the description of target and projectile atoms. Close-collision corrections to the Bethe stopping formula are treated on the basis of the transport cross section (8.4) much like the derivation of the Bloch correction by Lindhard and Sørensen (1996) but on the basis of a screening potential incorporating the adiabatic radius, thus providing an estimate of the Barkas-Andersen effect. Shell corrections for close collisions are incorporated via kinetic theory (5.7). Static projectile screening is characterized by a screening function composed of a sum of exponentials.

Table 9.1. Summary of theoretical schemes discussed in Sect. 9. Columns 3-10 list options for impact-parameter dependence (p), dependence on ion charge (q_1), Barkas-Andersen effect, shell correction, screening, projectile excitation and ionization (PE), Z_1 structure and Z_2 structure.

	v regime	p	q_1	Barkas-Andersen	shell	screen.	PE	Z_1	Z_2
UCA	$v_0 < v \ll c$	yes	yes	no	no	yes	yes[a]	no	yes
Binary	$v_0 \lesssim v$	yes	yes	yes	yes	yes	yes	no	yes
Nonlinear	$v \ll c$	no	yes	yes	yes	yes	no	yes	yes[b]
CKT	$Z_2 v_0 < v < Z_1 v_0$	no	yes	yes	no	yes	no	no	no

[a] Contribution from projectile ionization likely to be overestimated, see remarks in Sect. 6.7.

[b] Within limitations of the electron-gas model.

This model differs from the binary theory mainly by the use of an average-atom model for target and projectile (Maynard & al., 1996; Mabong & al., 1996) which avoids the use of empirical or semi-empirical oscillator strengths. The range of validity of the theory is asserted to be

$$Z_2 < \frac{v}{v_0} < Z_1. \tag{9.1}$$

Explicit applications of the theory have focused on hydrogen gas targets (Chabot & al., 1998; Gardès & al., 1998) but also include a study of the density effect in carbon (Maynard & al., 2000).

References

Arista, N. R. (2002). "Energy loss of heavy ions in solids: non-linear calculations for slow and swift ions," Nucl. Instrum. Methods B **195**, 91–105.

Arista, N. R. and Lifschitz, A. F. (1999). "Nonlinear calculation of stopping powers for protons and antiprotons in solids: the Barkas effect," Phys. Rev. A **59**, 2719–2722.

Arista, N. R. and Lifschitz, A. F. (2002). "Non-linear calculation of antiproton stopping powers at finite velocities using the extended Friedel sum rule," Nucl. Instrum. Methods B **193**, 8–14.

Azevedo, G. M., Grande, P. L., Behar, M., Dias, J. F. and Schiwietz, G. (2001). "Giant Barkas effect observed for light ion channeling in Si," Phys. Rev. Lett. **86**, 14821485.

Azevedo, G. M., Grande, P. L. and Schiwietz, G. (2000). "Impact-parameter dependent energy loss of screened ions," Nucl. Instrum. Methods B **164-165**, 203–211.

Bohr, N. (1913). "On the theory of the decrease of velocity of moving electrified particles on passing through matter," Philos. Mag. **25**, 10–31.

Calera-Rubio, J., Gras-Marti, A. and Arista, N. R. (1994). "Stopping power of low-velocity ions in solids - inhomogeneous electron-gas model," Nucl. Instrum. Methods B **93**, 137–141.

Chabot, M., Nectoux, M., Gardès, D., Maynard, G. and Deutsch, C. (1998). "Charge state dependence of the stopping power for chlorine ions interacting with a cold gas and a plasma (1.5 MeV/u)," Nucl. Instrum. Methods A **415**, 571–575.

Gardès, D., Chabot, M., Nectoux, M., Maynard, G., Deutsch, C. and Roudskoi, I. (1998). "Experimental study of stopping power for high Z ion in hydrogen," Nucl. Instrum. Methods A **415**, 698–702.

Grande, P. L., Araujo, L. L., DeAzevedo, G. M., Behar, M., Dias, J. F., DosSantos, J. H. R. and Schiwietz, G. (2002). "Energy loss under channeling conditions," Nucl. Instrum. Methods B 172–177.

Grande, P. L. and Schiwietz, G. (1998). "Impact-parameter dependence of the electronic energy loss of fast ions," Phys. Rev. A **58**, 3796–3801.

Grande, P. L. and Schiwietz, G. (2001). "CasP version 1.2," www.hmi.de/people/schiwietz/casp.html 451.

Grande, P. L. and Schiwietz, G. (2002). "The unitary convolution approximation for heavy ions," Nucl. Instrum. Methods B **195**, 55–63.

ICRU (2005). "Stopping of Heavy Ions," J. ICRU to appear.

Lifschitz, A. F. and Arista, N. (1998). "Velocity-dependent screening in metals," Phys. Rev. A **57**, 200–207.

Lindhard, J. and Sørensen, A. H. (1996). "On the relativistic theory of stopping of heavy ions," Phys. Rev. A **53**, 2443–2456.

Mabong, S., Maynard, G. and Katsonis, K. (1996). "Parametric potential for modelling of highly charged heavy ions," Laser and Particle Beams **14**, 575–586.

Maynard, G., Chabot, M. and Gardès, D. (2000). "Density effect and charge dependent stopping theories for heavy ions in the intermediate velocity regime," Nucl. Instrum. Methods B **164-165**, 139–146.

Maynard, G., Deutsch, C., Dimitriou, K., Katsonis, K. and Sarrazin, M. (2002a). "Evaluation of the energy deposition profile for swift heavy ions in dense plasmas," Nucl. Instrum. Methods B 188–215.

Maynard, G., Gardès, D., Chabot, M., Nectoux, M. and Deutsch, C. (1998a). "Effective stopping-power charges of swift heavy ions in gases," Nucl. Instrum. Methods B **146**, 88–94.

Maynard, G., Katsonis, K., Deutsch, C., Zwicknagel, G., Chabot, M. and Gardès, D. (2001a). "Modeling of swift heavy ions interaction with dense matter," Nucl. Instrum. Methods A **464**, 86–92.

Maynard, G., Katsonis, K. and Mabong, S. (1996). "Average atom model for swift heavy ions in dense matter," Nucl. Instrum. Methods B **107**, 51–55.

Maynard, G., Katsonis, K., Zwicknagel, G., Mabong, S., Chabot, M., Gardès, D. and Kurilenkov, Y. K. (1998b). "Nonlinear effects in stopping of partially ionized swift heavy ions," Nucl. Instrum. Methods A **415**, 687–692.

Maynard, G., Sarrazin, M., Katsonis, K. and Dimitriou, K. (2002b). "Quantum and classical stopping cross-sections of swift heavy ions derived from the evolution with time of the Wigner function," Nucl. Instrum. Methods B **193**, 20–25.

Maynard, G., Zwicknagel, G., Deutsch, C. and Katsonis, K. (2001b). "Diffusion-transport cross section and stopping power of swift heavy ions - art. no. 052903," Phys. Rev. A **63**, 052903-1–14.

Sigmund, P. and Schinner, A. (2000). "Binary stopping theory for swift heavy ions," Europ. Phys. J. D **12**, 425–434.

Sigmund, P. and Schinner, A. (2001a). "Binary theory of antiproton stopping," Europ. Phys. J. D **15**, 165–172.

Sigmund, P. and Schinner, A. (2001b). "Nonperturbative theory of charge-dependent heavy-ion stopping," Phys. Scr. **T92**, 222–224.

Sigmund, P. and Schinner, A. (2002a). "Binary theory of electronic stopping," Nucl. Instrum. Methods B **195**, 64–90.

Sigmund, P. and Schinner, A. (2002b). "Binary theory of light-ion stopping," Nucl. Instrum. Methods B **193**, 49–55.

10 Nuclear Stopping

10.1 Introductory Remarks

Nuclear stopping, i.e., energy transfer to recoiling nuclei, accounts for less than 0.05 % of all energy loss at projectile speeds above the orbital velocities of the majority of target electrons. Conversely the effect becomes dominating when the majority of electronic excitation channels is closed. Figure 4.2 indicates that the regime of dominating nuclear stopping typically lies within the velocity range $v < v_0$ except for rather heavy ions ($Z_1 \gtrsim 60$ and 40 for $Z_2 = 79$ and 6, respectively). Therefore, nuclear stopping is of minor importance in numerous situations considered in this monograph. On the other hand, unlike electronic collisions, recoil events are accompanied by angular deflection of the projectile that affects range and energy-deposition profiles as well as the analysis of stopping measurements.

The strong coupling between nuclear energy loss and angular deflection implies that nuclear stopping can only be measured with great difficulty in transmission measurements (Sidenius, 1974). Experimental evidence emerges more indirectly, mainly from measurements of angular scattering on gas targets and, to some extent, solid surfaces. Theory plays a key role in this area.

10.2 Binary Elastic Scattering

The conventional starting point for a theoretical treatment of recoil loss as well as angular scattering is the classical theory of binary elastic scattering[1] on a central force that is standard textbook material. This implies

1. decoupling of electronic from nuclear collisions,
2. negligible quantum effects,
3. existence of a spherically symmetric potential and

---

[1]The term 'elastic scattering' has different meanings in different disciplines. In the fields of x-ray, electron and neutron scattering on solids, elastic scattering means scattering without loss of translational energy, where the momentum change of the projectile is taken up by a macroscopic target. Scattering between heavy particles may be elastic with respect to the sum of the translational energies of the collision partners or, in other words, may be elastic in the center-of mass system.

4. negligible many-body effects.

A brief discussion of these items appears indicated.

Decoupling

Clearly, recoil and electronic losses are coupled. Indeed, electronic energy losses in individual collisions can be determined experimentally from measurements of the energies of particles scattered into a given direction[2] in conjunction with the law of momentum conservation. Both types of loss tend to decrease in magnitude with increasing impact parameter.

However, the maximum energy transfer from a heavy ion to a recoil atom far exceeds that to an electron. This causes recoil losses to dominate in close collisions. Conversely, at not-too-low projectile speeds, the effective range of electronic interaction exceeds that of the (screened) ion-atom interaction. This implies an approximate separation of nuclear and electronic energy loss in impact-parameter space as far as the contribution to the stopping cross section is concerned (Lindhard & al., 1963a,b).

As a rule of thumb, the significance of this type of correlation increases with decreasing energy down to the crossover of nuclear and electronic stopping. Quantitative estimates require knowledge of the dependence on impact-parameter of the mean nuclear energy loss and pertinent electronic-transition probabilities.

Quantum effects

The Bohr criterion for applicability of a classical-orbit description of binary scattering may be written in the form of (4.2) with an additional factor Z_2^2. This implies that the curves labelled 'classical' in Fig. 4.2 are shifted upward by a factor Z_2^2. Except for hydrogen targets this change is significant. It ensures a very wide regime of validity of classical-scattering theory.

Theoretical treatments based on classical descriptions of nuclear motion but quantal treatment of electronic excitation are common in atomic- and molecular-collision physics and go under the label 'semiclassical'.

Potential

Classical binary-scattering theory makes explicit use of angular-momentum conservation. Therefore the assumption of a central force is essential. Screened-Coulomb interaction potentials of the type of

$$V(R) = \frac{Z_1 Z_2 e^2}{4\pi\epsilon_0 R} u\left(\frac{R}{a}\right) \tag{10.1}$$

---

[2]The same is true for inelastic losses in nuclear reactions.

are most common. Here R is the distance between the interacting nuclei, u some screening function and a a screening radius. Both u and a depend on Z_1, Z_2 and the charge state q_1. These dependencies are frequently expressed in terms of Thomas-Fermi type scaling relations to be discussed below.

Many-body effects

It is appropriate to distinguish between many-body collisions and many-body potentials.

Many-body collisions occur in dense (solid) media at energies so low that the dimensions of a scattering trajectory are of the order of or exceed the internuclear distance in the medium. This is typically the case at particle energies well below 1 keV. In practice, collision problems in this energy range are treated by molecular-dynamics or alternative types of computer simulation (Eckstein, 1991).

Many-body potentials describe the interaction between two particles embedded in a medium. The medium is typically characterized as a Fermi gas of constant density, and the interaction between embedded atoms takes into account the relaxation of the electron gas around the intruders. This type of interaction, called 'embedded medium' (Jacobsen & al., 1987) or 'embedded-atom' (Daw and Baskes, 1984) potentials, is elastic but non-binary and hence needs to be treated by some molecular-dynamics computer code. The theoretical formalism behind the embedded-medium theory is closely related to the density-functional approach that also enters the theory of low-speed electronic stopping discussed in Sect. 8.6.

An extensive survey on interatomic potentials in radiation physics has been given by Dedkov (1995).

10.3 Scaling Properties

For elastic scattering on a central potential of the form of (10.1), a rigorous scaling relation follows from conservation laws and classical scattering theory (Lindhard & al., 1968):

$$\theta = \theta\left(\frac{p}{a}, \epsilon\right), \tag{10.2}$$

where θ is the center-of-mass scattering angle, p the impact parameter, and ϵ a dimensionless parameter[3] proportional to the energy defined as

$$\epsilon = \frac{4\pi\epsilon_0 M_0 v^2 a}{2 Z_1 Z_2 e^2}, \tag{10.3}$$

---

[3]The symbol ϵ occurs with different meanings in the stopping literature, such as beam energy, stopping force, stopping cross section and stopping number. In the present report, ϵ denotes consistently the quantity defined by (10.3).

where $M_0 = M_1 M_2/(M_1 + M_2)$ is the reduced mass.

The energy transfer w to a target atom initially at rest is known to be given by

$$w = \gamma E \sin^2 \frac{\theta}{2},$$ (10.4)

with[4] $\gamma = 4M_1 M_2/(M_1 + M_2)^2$. From this follows the differential cross section

$$d\sigma(w) = \pi a^2 g\left(\epsilon, \frac{w}{\gamma E}\right) \frac{dw}{\gamma E},$$ (10.5)

with the function g determined by the screening function u, and the stopping force

$$-\frac{dE}{d\ell} = nS = \int_0^{\gamma E} w\, d\sigma(w) = N\pi a^2 \gamma E s(\epsilon),$$ (10.6)

where $s(\epsilon) = \int_0^1 dt\, t g(t)$ is a function determined by u. The latter relation is commonly written in the form

$$-\frac{d\epsilon}{d\rho} = s(\epsilon),$$ (10.7)

where ρ is a dimensionless measure of pathlength

$$\rho = n\pi a^2 \gamma \ell.$$ (10.8)

While these relations are quite general, they become particularly useful when the screening function u and the screening radius a themselves show simple scaling properties with Z_1 and Z_2. Frequently, universal screening functions are adopted (Bohr, 1948; Firsov, 1957; Lindhard & al., 1968; Ziegler & al., 1985) and screening radii like those shown in Table 10.1. If u is a universal screening function, also $s(\epsilon)$ becomes a universal function.

Table 10.1. Common scaling relations for screening radius a.

Bohr (1948):	$a_0 \left(Z_1^{2/3} + Z_2^{2/3}\right)^{-1/2}$
Lindhard & al. (1968):	$0.885 a_0 \left(Z_1^{2/3} + Z_2^{2/3}\right)^{-1/2}$
Firsov (1957):	$0.885 a_0 \left(Z_1^{1/2} + Z_2^{1/2}\right)^{-2/3}$
Ziegler & al. (1985):	$0.885 a_0 \left(Z_1^{0.23} + Z_2^{0.23}\right)^{-1}$

---

[4]For the notation of γ cf. footnote 2 on page 7.

In addition to these exact scaling properties, approximate scaling has been demonstrated for repulsive screening functions by Lindhard & al. (1968), where the dependence on two variables in (10.5) is replaced by one variable

$$\eta = \epsilon \sin \frac{\theta}{2}, \tag{10.9}$$

so that

$$d\sigma(w) = \pi a^2 \frac{d\eta}{\eta^2} f(\eta), \tag{10.10}$$

$$s(\epsilon) = \frac{1}{\epsilon} \int_0^\epsilon d\eta f(\eta), \tag{10.11}$$

and $f(\eta)$ is determined by the adopted screening function.

Table 10.2. Coefficients entering scaling function $f(\eta)$ for differential scattering cross section, (10.12).

Screening function u	m	q	λ
Thomas-Fermi	0.333	0.667	1.309
Thomas-Fermi-Sommerfeld	0.311	0.588	1.70
Lenz-Jensen	0.191	0.512	2.92
Molière	0.216	0.570	2.37
Bohr	0.103	0.570	2.37

Convenient analytical approximations for $f(\eta)$ were given by Winterbon & al. (1970) and Winterbon (1972),

$$f(\eta) \simeq \frac{\lambda \eta^{1-2m}}{\left(1 + \left[2\lambda \eta^{2(1-m)}\right]^q\right)^{1/q}} \tag{10.12}$$

with coefficients m, q and λ given in Table 10.2.

Plots of the functions $f(\eta)$ and $s(\epsilon)$ for these potentials are shown in Figs. 10.1 and 10.2. There is general agreement that the Thomas-Fermi screening function overestimates the potential at large distances and hence the stopping cross sections at low energies, and that the opposite is the case for the Bohr screening function. The difference between Lenz-Jensen and Molière, although noticeable especially in the low-energy range, reflects the inherent uncertainty of this type of potentials. Figure 10.2 also includes the reduced stopping cross section in the form favored by Ziegler & al. (1985) – denoted by ZBL – which may be written in the form

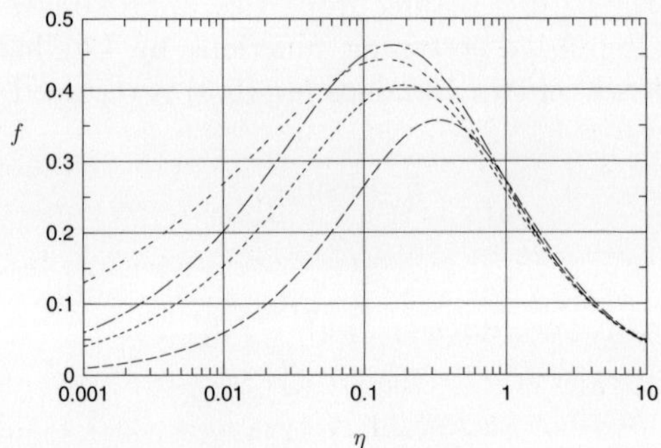

Fig. 10.1. Function $f(\eta)$, (10.12), determining scaled differential scattering cross section for Thomas-Fermi, Molière, Lenz-Jensen and Bohr potentials (top to bottom at the left edge of the graph). Screening functions specified in Table 10.2.

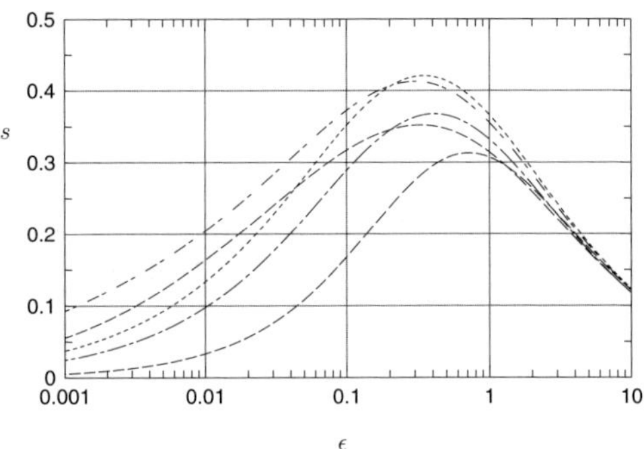

Fig. 10.2. Function $s(\epsilon)$, (10.11), determining scaled nuclear-stopping cross section for Thomas-Fermi, ZBL, Molière, Lenz-Jensen and Bohr potential (top to bottom on left edge of the graph). Screening functions specified in Table 10.2.

$$s(\epsilon) = \frac{\ln(1 + a\epsilon)}{2\left(\epsilon + b\epsilon^c + d\sqrt{\epsilon}\right)},\tag{10.13}$$

with

$$a = 1.1383; b = 0.01321; c = 0.21226; d = 0.19593.$$

An additional source of uncertainty lies in the choice of screening radius. Common choices are shown in Table 10.1. The choice of Ziegler & al. (1985), which differs noticeably from the others, was generated by an averaging process involving numerically computed interaction potentials for a very large number of atom-atom pairs. These potentials were computed on the basis of the Thomas-Fermi method with atomic charge distributions as the main input. The value of a screening radius extracted from such potential functions clearly depends on the weights assigned to different regimes of interatomic distance. The procedure by Bohr (1948), followed also by Lindhard & al. (1968) and Firsov (1957), refers to the regime of weak screening, $R \lesssim a$, while the averaging process adopted by Ziegler & al. (1985) involves the range $R \lesssim 25a$, i.e., far beyond the range of validity of the Thomas-Fermi model.

Although interaction potentials unquestionably depend on the charge states of the collision partners, this dependence is usually neglected, the main reason being that in the low-energy region, where the outer parts of the interaction potential play a most significant role, collision partners are most often neutral. For distant collisions, a simple way to approximately account for the charge state of the ion is to replace Z_1 by q_1 in the expression for the screening radius (Amsel & al., 2003).

10.4 Power-Law Cross Sections

According to (10.12), the function $f(\eta)$ approaches power form at low arguments,

$$f(\eta) \simeq \lambda \eta^{1-2m} \qquad \text{for } \eta \ll 0.1. \tag{10.14}$$

At large values of η, on the other hand, all curves approach a common asymptotic form

$$f(\eta) \sim \frac{1}{2\eta} \qquad \text{for } \eta \gg 1 \tag{10.15}$$

reflecting Rutherford scattering and $m = 1$. In the intermediate regime, a constant equivalent with the choice $m = 1/2$ may be employed in rough estimates. The function adopted by Bohr (1948) assumes $m = 1/2$ and $\lambda = 0.327$.

Stopping cross sections following from (10.14) take the form

$$s(\epsilon) = \frac{\lambda}{2(1-m)} \epsilon^{1-2m}. \tag{10.16}$$

In particular, for $m = 1/2$, a constant stopping cross section $s = 0.327$ was found (Bohr, 1948).

Power-law cross sections are convenient for rough analytical estimates and have been employed extensively in the theory of radiation effects. Their general form is

$$d\sigma(E, w) = C \frac{dw}{E^m w^{1+m}}, \tag{10.17}$$

with

$$C = \frac{\pi}{2}\lambda a^2 \left(\frac{M_1}{M_2}\right)^m \left(\frac{2Z_1 Z_2 e^2}{4\pi\epsilon_0 a}\right)^{2m},\tag{10.18}$$

and parameters m, λ depending on the range of η and/or ϵ where the expression is going to be used.

Power-law cross sections were originally derived by approximating the screening function u by an inverse-power form (Bohr, 1948; Lindhard & al., 1968).

10.5 Concluding Remarks

It is emphasized that the treatment of nuclear stopping offered in this section addresses situations for which the effect is of minor importance.

Moreover, using scaling relations is a matter of convenience and by no means necessary. Computation of classical-scattering integrals is a routine matter which, at least for purely repulsive interaction potentials, does not require excessive computation times. Therefore, when a good estimate of an interaction potential is available, a reliable scattering law ought to be generated by direct integration rather than by scaling.

Finally it is emphasized that, dependent on the experimental setup, only a restricted nuclear-stopping cross section might contribute to the measured energy loss. For the frequently used transmission technique, this aspect will be considered in Sect. 15.

References

Amsel, G., Battistig, G. and L'Hoir, A. (2003). "Small angle multiple scattering of fast ions, physics, stochastic theory and numerical calculations," Nucl. Instrum. Methods B **201**, 325–388, URL www.mfm.kfki.hu/ms.

Bohr, N. (1948). "The penetration of atomic particles through matter," Mat. Fys. Medd. Dan. Vid. Selsk. **18 no. 8**, 1–144.

Daw, M. S. and Baskes, M. I. (1984). "Embedded-atom method – derivation and application to impurities, surfaces, and other defects in metals," Phys. Rev. B **29**, 6443–6453.

Dedkov, G. V. (1995). "The interatomic interaction potentials in radiation physics," phys. stat. sol. A **149**, 453–514.

Eckstein, W. (1991). *Computer simulation of ion-solid interactions* (Springer-Verlag, Berlin).

Firsov, O. B. (1957). "Interaction energy of atoms for small nuclear separations," Zh. Eksp. Teor. Fiz. **32**, 1464–1469, [English translation: Sov. Phys. JETP **5**, 1192-1196 (1957)].

Jacobsen, K. W., Nørskov, J. K. and Puska, M. J. (1987). "Interatomic interactions in the effective-medium theory," Phys. Rev. B **35**, 74237442.

Lindhard, J., Nielsen, V. and Scharff, M. (1968). "Approximation method in classical scattering by screened coulomb fields," Mat. Fys. Medd. Dan. Vid. Selsk. **36 no. 10**, 1–32.

Lindhard, J., Nielsen, V., Scharff, M. and Thomsen, P. V. (1963a). "Integral equations governing radiation effects," Mat. Fys. Medd. Dan. Vid. Selsk. **33 no. 10**, 1.

Lindhard, J., Scharff, M. and Schiøtt, H. E. (1963b). "Range concepts and heavy ion ranges," Mat. Fys. Medd. Dan. Vid. Selsk. **33 no. 14**, 1.

Sidenius, G. (1974). "Systematic stopping cross section measurements with low energy ions in gases," Mat. Fys. Medd. Dan. Vid. Selsk. **39 no. 4**, 1–32.

Winterbon, K. B. (1972). "Heavy-ion range profiles and associated damage distributions," Radiat. Eff. **13**, 215–226.

Winterbon, K. B., Sigmund, P. and Sanders, J. B. (1970). "Spatial distribution of energy deposited by atomic particles in elastic collisions," Mat. Fys. Medd. Dan. Vid. Selsk. **37 no. 14**, 1–73.

Ziegler, J. F., Biersack, J. P. and Littmark, U. (1985). "The stopping and range of ions in solids," J. F. Ziegler, ed., *The Stopping and Ranges of Ions in Matter*, volume 1 of *The Stopping and Ranges of Ions in Matter*, 1–319 (Pergamon, New York).

11 Related Processes

11.1 Nuclear Reactions

Cross sections for nuclear reactions leading to a change in mass and/or identity of the projectile are much smaller than for atomic processes except for relativistic heavy ions with more than several hundred MeV/u, at which energies they may become comparable. This is illustrated in Fig. 11.1, where the nuclear reaction probability is calculated for different ions in carbon, aluminum and lead. The total nuclear-reaction cross section applied in this calculation consists of a pure nuclear part calculated according to Shen $\mathcal{E}$ al. (1989) and a component due to electromagnetic dissociation (Bertulani and Baur, 1988), which contributes mainly for large Z_1 and/or Z_2. The comparison demonstrates that for lighter projectiles and lighter stopping media the losses due to nuclear reactions are severe. This is the reason why light materials like beryllium are preferred as production targets in high-energy fragmentation reactions. Conversely, for electromagnetic dissociation of relativistic projectiles, lead targets are used successfully in studies and applications of exotic nuclei (Geissel, 1997).

11.2 Channeling, and Stopping at Surfaces

Although energy loss under channeling conditions is a problem of considerable complexity, the number of application areas is narrower than that of stopping in random media. Therefore the present section is intended mainly to give a brief survey of complicating features. For a general introduction to channeling the reader is referred to a review by Gemmell (1974).

An ion beam penetrating through a medium with a regular crystalline structure can be considered as being split into two components, a random and an aligned beam (Lindhard, 1965). The aligned beam is guided by the collective action of closely spaced atoms making up rows or planes of atoms and hence avoids close encounters with the target nuclei. This gives rise to dramatic reductions in the rates of wide-angle Rutherford scattering, nuclear reactions, inner-shell excitation and multiple angular scattering. Energy loss is also reduced, but that effect is less dramatic because of the long range of

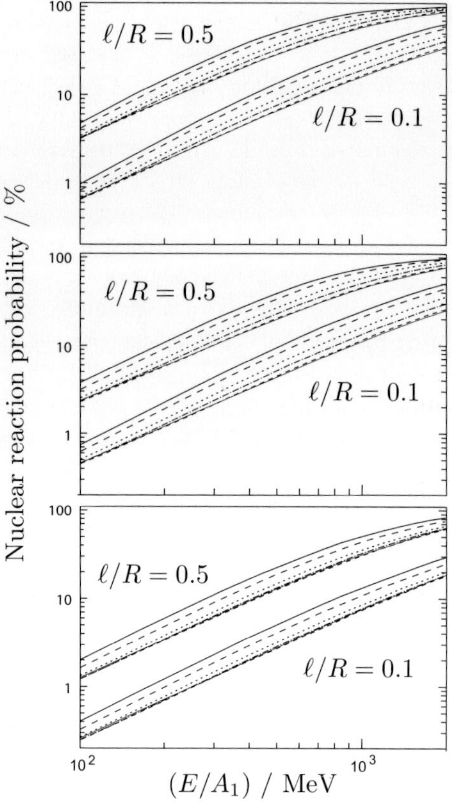

Fig. 11.1. Nuclear reaction probability during the slowing down of heavy ions in carbon, aluminum, and lead (top to bottom graph). ℓ is the path length and R the total range. Lines for ^{20}Ne, ^{40}Ar, ^{86}Kr, ^{132}Xe, ^{208}Pb and ^{238}U (top to bottom in each group). From Geissel & al. (2002).

electronic interactions in the high-energy range where channeling phenomena are most pronounced.

Theoretical descriptions of energy loss in channeling are traditionally based on electron-gas models (Lindhard, 1965), the underlying reason being the dominance of outer target shells in the energy-loss process. The equipartion rule of Lindhard and Winther (1964) – valid for light ions penetrating through a homogeneous Fermi gas – suggests reduction by roughly a factor of two of the mean energy loss at not too low projectile speeds. An extensive body of literature is devoted to this topic, based on more or less sophisticated models of homogeneous and inhomogeneous Fermi gases.

More quantitative estimates are now possible on the basis of impact-parameter-dependent energy-loss functions for covalent and ionic materials, supplemented by electron-gas estimates for conduction electrons in metals. This requires impact-parameter distributions that, typically, are found from binary-collision or molecular-dynamics simulation of pertinent trajectories. Energy-loss functions $w(v, p)$ dependent on impact parameter may be determined from schemes discussed in Sect. 9, especially the binary theory and the unitary-convolution approximation. Initial steps have been made (Sigmund and Schinner, 2001; Azevedo & al., 2001), but a considerable effort lies ahead in analysing a large body of unexplained experimental data.

In addition to target thickness and projectile speed, the energy-loss spectrum for a given beam depends on the orientation of the target and the angle between the beam and the closest crystal chain or plane. For heavy ions, charge equilibration has been found to be dramatically affected (Datz & al., 1972). This provides the possibility of measurements with ions in 'frozen charge states' (Datz & al., 1977; Golovchenko & al., 1981).

Separation of measured energy-loss spectra into a random and a channeled fraction is frequently possible. In addition, the energy loss of 'best-channeled' particles can occasionally be identified by extrapolation of the low-loss or high-penetration tail. This is in contrast to random slowing-down in which Poisson statistics does not allow this distinction.

With appropriate modifications the present considerations also apply to measurements of scattering and energy loss of well-collimated beams on flat surfaces (Kimura & al., 1987; Winter, 2002).

References

Azevedo, G. M., Grande, P. L., Behar, M., Dias, J. F. and Schiwietz, G. (2001). "Giant Barkas effect observed for light ion channeling in Si," Phys. Rev. Lett. **86**, 14821485.

Bertulani, C. A. and Baur, G. (1988). "Electromagnetic processes in relativistic heavy-ion collisions," Phys. Rep. **163**, 299–408.

Datz, S., DelCampo, J. G., Dittner, P. F., Miller, P. D. and Biggerstaff, J. A. (1977). "Higher-order Z_1 effects and effects of screening by bound K electrons on the electronic stopping of channeled ions," Phys. Rev. Lett. **38**, 1145–1148.

Datz, S., Martin, F. W., Moak, C. D., Appleton, B. R. and Bridwell, L. B. (1972). "Charge-changing collisions of channeled oxygen ions in gold," Radiat. Eff. **12**, 163–169.

Geissel, H. (1997). "Relativistische exotische Kerne als Projektilstrahlen – Neue Perspektiven zum Studium der Kerneigenschaften," GSI Report **97-03**, 1, habilitation thesis.

Geissel, H., Weick, H., Scheidenberger, C., Bimbot, R. and Gardès, D. (2002). "Experimental studies of heavy-ion slowing down in matter," Nucl. Instrum. Methods B **195**, 3–54.

Gemmell, D. S. (1974). "Channeling and related effects in motion of charged particles through crystals," Rev. Mod. Phys. **46**, 129–227.

Golovchenko, J. A., Goland, A. N., Rosner, J. S., Thorn, C. E., Wegner, H. E., Knudsen, H. and Moak, C. D. (1981). "Charge state dependence of channeled ion energy loss," Phys. Rev. B **23**, 957–966.

Kimura, K., Hasegawa, M. and Mannami, M. (1987). "Energy loss of MeV light ions specularly reflected from a SnTe(001) surface," Phys. Rev. B **36**, 7–12.

Lindhard, J. (1965). "Influence of crystal lattice on motion of energetic charged particles," Mat. Fys. Medd. Dan. Vid. Selsk. **34 no. 14**, 1–64.

Lindhard, J. and Winther, A. (1964). "Stopping power of electron gas and equipartition rule," Mat. Fys. Medd. Dan. Vid. Selsk. **34 no. 4**, 1–22.

Shen, W., Wang, B., Feng, J., Zhan, W., Zhu, Y. and Feng, E. (1989). "Total reaction cross section for heavy-ion collisions and its relation to the neutron excess degree of freedom," Nucl. Instrum. Methods A **282**, 247.

Sigmund, P. and Schinner, A. (2001). "Resolution of the frozen-charge paradox in stopping of channeled heavy ions," Phys. Rev. Lett. **86**, 1486–1489.

Winter, H. (2002). "Collisions of Atoms and Ions with Surfaces under Grazing Incidence," Phys. Repts. **367**, 387–582.

12 Statistics of Particle Penetration

12.1 Qualitative Survey

The importance of statistical considerations in the physics of particle penetration has been emphasized in particular by Bohr (1948). An extensive review of the subject matter — including outline for work to be done — was given by Sigmund (1991). The present section serves to summarize central results that affect the analysis of energy-loss measurements and the relation between measured and calculated stopping parameters.

It is useful to distinguish between measurements on

- *thin targets* in which a beam particle can interact with only a small number of target atoms,
- *thick targets* in which the number of interactions is large enough to make fluctuations small, and
- *very thick targets* in which the number of interactions is so large that beam particles lose a significant fraction of their energy, thus giving rise to nonnegligible variations in pertinent cross sections.

For thin targets, the energy-loss spectrum and angular distribution of an initially monochromatic and well-collimated beam are modified in accordance with the cross section for energy loss and angular deflection, respectively. This implies a pronounced peak around zero energy loss and a tail extending toward the maximum energy loss or scattering angle in a single collision.

For thick targets the energy-loss spectrum has a trend toward gaussian shape centered around the average energy loss, while the angular distribution tends toward a gaussian centered around the initial beam direction.

For very thick targets pronounced changes in pertinent cross sections cause deviations from gaussian shape that need special consideration.

These differences are highly visible for charged-particle interactions since cross sections for Coulomb scattering are power-like and thus differ dramatically from gaussian shape. This, at the same time, makes it possible to operationally distinguish between the thin- and thick-target regime for a given region in the (Z_1, Z_2, v) parameter space by mere inspection of measured or calculated spectra.

Although Coulomb scattering is the predominant factor in collision statistics for light penetrating particles, charge exchange enters as a complicating phenomenon in case of heavier ions. This influence is twofold:

- Cross sections for energy loss and angular deflection may depend on charge state, and
- Charge-exchange events may themselves give rise to significant energy loss and/or angular scattering.

In addition to these intrinsic and unavoidable sources of fluctuations, observed fluctuation phenomena are affected by experimental factors like target nonuniformity, beam instabilities and counting statistics.

12.2 Stripped Ions

Fluctuation phenomena are well-investigated for penetrating point charges, i.e., for ions with velocities $v \gg Z_1^{2/3} v_0$, cf. Fig. 4.2. Energy-loss statistics for both thin and thick targets – but not for very thick targets – is governed by the Bothe-Landau formula (Landau, 1944) for the energy-loss spectrum,

$$F(\Delta E, \ell)\mathrm{d}(\Delta E) = \frac{\mathrm{d}(\Delta E)}{2\pi} \int_{-\infty}^{\infty} \mathrm{d}s\, e^{is\Delta E - n\ell\sigma(s)}, \qquad (12.1)$$

for random slowing down, i.e., uniform distribution of scattering events in space and time. Here, s is a variable in Fourier space and

$\Delta E = $ total energy loss
$\ell = $ path length
$n = $ number of target atoms per volume
$\sigma(s) = \int \mathrm{d}\sigma(w) \left(1 - e^{-isw}\right) = $ transport cross section
$w = $ energy loss in individual event
$\mathrm{d}\sigma(w) = $ differential cross section per target atom for energy loss in interval $(w, \mathrm{d}w)$.

One may derive (2.8) – (2.14) from (12.1) by taking first- and second-order moments over ΔE. The main question of interest is how these averages relate to peak value and halfwidth of the spectrum, which are often more accessible to measurement.

When ℓ is large, all penetrating particles experience large energy losses so that the integral (12.1) receives contributions mainly from small values of s. Expansion of $\sigma(s)$ up to second order in s and subsequent integration then yields the gaussian

$$F(\Delta E, \ell) \simeq \frac{1}{\sqrt{2\pi n\ell W}} e^{-(\Delta E - n\ell S)^2 / n\ell W}, \qquad (12.2)$$

with the stopping cross section $S = \int w\,\mathrm{d}\sigma(w)$ and straggling parameter $W = \int w^2 \mathrm{d}\sigma(w)$.

For thin targets Landau (1944) presented the solution

$$F(\Delta E, \ell) = \frac{w_{\max}}{\Omega_B^2} \frac{1}{2\pi i} \times \int_{c-i\infty}^{c+i\infty} \mathrm{d}s\, e^{s\ln s + \lambda s} \quad (c > 0), \qquad (12.3)$$

$$w_{\max} = 2mv^2/(1 - v^2/c^2),$$

$$\Omega_B^2 = 4\pi Z_1^2 Z_2 e^4 n\ell,$$

which expresses the spectrum by a single variable

$$\lambda = \frac{w_{\max}(\Delta E - \langle \Delta E \rangle)}{\Omega_B^2} - \ln \frac{\Omega_B^2}{w_{\max}^2} - 1 + \gamma \qquad (12.4)$$

$$\gamma = 0.5772 = \text{Euler's constant.}$$

Note that unlike (12.2), (12.3) implies Coulomb scattering specifically.

Equation (12.4) suggests that the peak position does not normally lie at $\Delta E = \langle \Delta E \rangle$ and is governed by the ratio $\Omega_B^2/w_{\max}^2$.

The range of validity of these approximations is given by $\Omega_B/w_{\max} \gg 1$ for the gaussian and $\Omega_B/w_{\max} \ll 1$ for the Landau solution. Attempts to reduce the large intermediate regime where neither approximation is applicable include the schemes of Vavilov (1957), Symon (1948) and Sigmund and Winterbon (1985) to expand the high-ℓ regime downward and of Lindhard (1985) to expand the Landau regime upward.

Most successful has been a dual approach by Glazov (2000) that invokes an extension of the Landau scheme toward larger thicknesses and of the steepest-descent method (Sigmund and Winterbon, 1985) toward smaller thicknesses, with a comfortable overlap regime at sufficiently high energies. An example is shown in Fig. 12.1.

Extraction of stopping parameters from measured spectra can be simplified if the mean energy loss $\langle \Delta E \rangle$ is close to the peak value ΔE_{peak}, and if the relation between standard deviation and halfwidth is the one familiar from the gaussian. As long as deviations are small, the following relations, derived by Sigmund and Winterbon (1985) for the large-thickness limit may be useful,

$$\Delta E_{\text{peak}} = \langle \Delta E \rangle - \frac{mv^2}{2} + \mathcal{O}\left\{\frac{1}{n\ell}\right\}, \qquad (12.5)$$

and

$$\Delta E_{\pm 1/2} = \Delta E_{\text{peak}} \pm 1.177\Omega_B + 0.231mv^2 \mp 0.098\frac{(mv^2)^2}{\Omega_B} + \mathcal{O}\left\{\frac{1}{n\ell}\right\}, \quad (12.6)$$

where $\mathcal{O}\{1/n\ell\}$ indicates the order in $n\ell$ of the leading neglected term.

The range of applicability of the above tools can be extended into the range of very thick targets by replacing the initial energy E by an effective energy

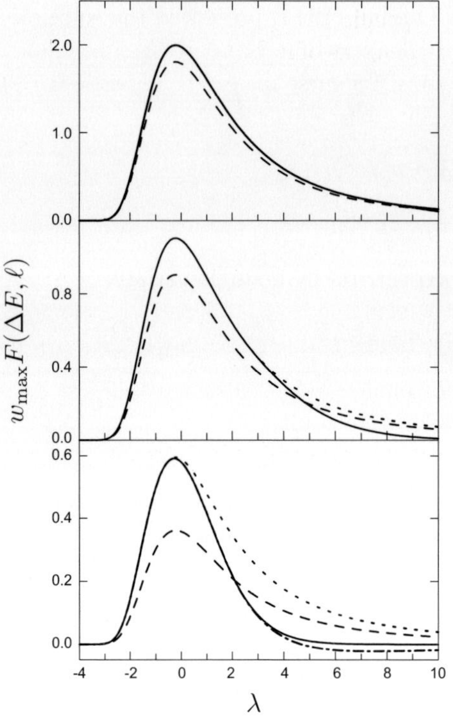

Fig. 12.1. Approximations to the energy-loss spectrum of a swift point charge in dimensionless units. Solid line: Numerical evaluation of (12.1); dashed line: (12.3); dotted and dash-dotted lines refer to successive approximations to an expanded Landau approximation. From Glazov (2000).

$$E_{\text{eff}} = E - \langle \Delta E \rangle / 2, \tag{12.7}$$

or by replacing the true pathlength ℓ by an effective pathlength

$$\ell_{\text{eff}} = \ell \left(1 + \frac{\alpha}{2} \frac{\langle \Delta E \rangle}{E} \right), \tag{12.8}$$

where α is a numerical coefficient characterizing the energy dependence of the transport cross section ($\sigma(s) \propto E^{-\alpha}$) (Sigmund, 1991). A useful relation for straggling was derived by Symon (1948),

$$\Omega^2 = \left(NS(E_1) \right)^2 \int_{E_1}^{E_0} dE' \frac{NW(E')}{\left(NS(E') \right)^3}, \tag{12.9}$$

where E_1 is the exit energy.

Depending on the specific system and desired accuracy, such schemes may be adequate up to pathlengths of half the range. For larger thicknesses and/or high-accuracy estimates, recourse has been made to tools from range theory (Symon, 1948; Tschalär, 1968).

12.3 Partially-Stripped Ions

For partially-stripped ions the above formalism is applicable as it stands only as long as charge exchange is insignificant, i.e., for *frozen charges*. In the presence of charge exchange several complications arise:

- The effective *collisional stopping force* becomes a weighted mean of *frozen-charge stopping forces*,
- The variation in the frozen-charge stopping force is a source of straggling (*charge-exchange straggling*),
- Energy loss in electron capture and loss is a separate contribution to the total energy loss,
- Electronic processes on the projectile such as excitation, deexcitation or Auger decay may contribute to the energy balance, affect pertinent cross sections and occur on a separate time scale that is only indirectly related to the projectile speed.

A formalism that, slightly modified, can allow for the above features, was presented by Winterbon (1977). It is based on a linear transport equation and hence applicable to the entire range of target thicknesses. The formalism proposed by Sigmund (1992) is equivalent in physical content, but being based on an extension of (12.1) it relates more directly to well-known standard results for point charges. Moreover, notation was chosen such as to allow for explicit incorporation of all effects mentioned above to the extent that pertinent atomic parameters are available.

The scheme operates with states $I, J \ldots$ of the ion, which may denote charge and/or excitation states. One then introduces an energy-loss spectrum $F_{IJ}(\Delta E, \ell)\mathrm{d}(\Delta E)$, where I denotes the initial state and J the state after pathlength ℓ. The matrix $\mathbf{F}(\Delta E, \ell)$ obeys the generalized Bothe-Landau formula (Sigmund, 1991, 1992)

$$\mathbf{F}(\Delta E, \ell) = \frac{1}{2\pi} \int_{-\infty}^{\infty} \mathrm{d}s\, \mathrm{e}^{is\Delta E + n\ell \mathbf{Q} - n\ell \boldsymbol{\sigma}(s)}, \qquad (12.10)$$

where $\mathbf{Q}$, $\boldsymbol{\sigma}$ and $\boldsymbol{\sigma}(s)$ represent matrices with elements Q_{IJ} and σ_{IJ}, respectively,

$$Q_{IJ} = \sigma_{IJ} - \delta_{IJ} \sum_L \sigma_{IL} \qquad (12.11)$$

$$\sigma_{IJ} = \int d\sigma_{IJ}(w) \qquad (12.12)$$

$$\boldsymbol{\sigma}(s) = \int d\sigma_{IJ}(w)\left(1 - e^{-isw}\right), \qquad (12.13)$$

and $d\sigma_{IJ}(w)$ the differential cross section for energy loss (w, dw) in a collision with the ion in incident and final states I and J, respectively. In this notation only collision-induced processes are assumed active. In the presence of spontaneous processes such as Auger decay a notation based on transition rates is more convenient (Sigmund, 1992).

Integration of (12.10) over the energy loss yields the probability

$$P_{IJ}(\ell) = \left(e^{n\ell \mathbf{Q}}\right)_{IJ} \qquad (12.14)$$

for an ion in the initial state I to be in state J after a pathlength ℓ. Equation (12.14) is a compact solution of the familiar rate equations that are normally used to describe the approach to charge equilibrium (Allison, 1958).

The occurrence of a matrix in an exponential is a complication which can be overcome by standard mathematical tools such as eigenvalue expansion for the approach to equilibrium (Sigmund, 1992; Glazov and Sigmund, 1997) or Taylor expansion when the number of charge exchanges is small (Sigmund, 1994; Glazov, 2002).

In the present context, prime quantities are stopping force and straggling in charge equilibrium and information about the approach to equilibrium from a given initial charge state. Such information is given in the form of asymptotic expansions in terms of $1/n\ell$ similar to (12.5) and (12.6). If only the equilibrium value and the first correction term are given, the latter produces an intercept for the straight lines $\langle \Delta E \rangle$ or Ω^2 versus ℓ.

General expressions for equilibrium and intercept were presented by Sigmund (1992) and Närmann and Sigmund (1994). For equilibrium stopping, (6.2) is obtained. Straggling is made up by a term analogous to (6.2) accounting for collisional straggling and another one that vanishes in the absence of charge exchange. The latter can be written in a particularly transparent way for the specific case of a two-state system for which (Sigmund, 1992)

$$\Omega^2_{\text{chex}} = 2n\ell \frac{(S_1 - S_2)}{(\sigma_{12} + \sigma_{21})^3} \times \left[\sigma_{12}\sigma_{21}\left(S_{11} - S_{22}\right) + \left(\sigma_{12}^2 S_{21} - \sigma_{21}^2 S_{12}\right)\right],$$

$$(12.15)$$

$$S_{IJ} = \int w d\sigma_{IJ}(w),$$
$$S_I = \sum_J S_{IJ}.$$

If energy loss in charge exchange is negligible, only the first term in the square brackets remains, and Ω_{chex} reduces to the well-known expression (Efken $\&$ al., 1975)

$$\Omega^2_{\text{chex}} \simeq 2n\ell \frac{(S_1 - S_2)^2}{(\sigma_{12} + \sigma_{21})^3} \sigma_{12}\sigma_{21}. \tag{12.16}$$

For a three-state system, energy loss in charge exchange being neglected, the corresponding expression reads

$$\Omega^2_{\text{chex}} = \frac{n\ell}{\beta^3} \sum_{JK} (S_J - S_K)^2 \mu_J (\alpha\mu_K - \beta\sigma_{JK}) \tag{12.17}$$

$\alpha = \sum_{KL} \sigma_{KL}$
$\mu_1 = \sigma_{32}\sigma_{21} + \sigma_{23}\sigma_{31} + \sigma_{21}\sigma_{31}$
$\mu_2, \mu_3 = $ cyclical permutations
$\beta = \sum_J \mu_J.$

12.4 Transport Equations and Simulation

To the extent that the assumption of random slowing down is satisfied, linear transport equations constitute a convenient and accurate analytical tool for problems that cannot be handled efficiently on the basis of the Bothe-Landau formula (12.1) or its expanded version (12.10). This includes in particular

- very thick targets allowing substantial energy loss, including complete slowing down,
- problems involving secondary particles such as recoil atoms and ejected electrons,
- complex geometries, including composite targets.

The original derivation of (12.1) by Landau (1944) started off from

$$\frac{\partial}{\partial \ell} F(E, \ell) = n \int dw \left(K(E + w, w)F(E + w, \ell) - K(E, w)F(E, \ell) \right), \tag{12.18}$$

where $F(E, \ell)$ is the *energy spectrum* at pathlength ℓ and $K(E, w) = d\sigma(w)/dw$ at the energy E. Equation (12.18) is a *forward transport equation* for the energy spectrum versus pathlength. Additional spectral variables like direction of motion and charge state are easily incorporated. This complex has been summarized by Sigmund (1991), where also backward equations are discussed.

For thin and thick targets the most appropriate way to solve transport equations is via the Bothe-Landau scheme. For targets that are too thick to allow this procedure, the traditional way of solving the equations goes over moments either of the energy or the pathlength. Reconstruction of a spectrum from its moments is notoriously difficult (Symon, 1948; Winterbon & al., 1970). With easy access to present-day computers this route does not any longer reflect the state-of-the-art. Numerical solutions of transport equations or direct Monte Carlo simulation may easily be superior in both efficiency and accuracy.

12.5 Non-Poisson Statistics

When the probability for collision processes is distributed uniformly in space and time, their frequency distribution is governed by Poisson statistics. Two configurations are known where pronounced deviations from Poisson behavior can occur:

- In a dense medium atoms are arranged in close packing rather than at random. This means that processes governed by large cross sections, i.e. with a mean free path not much larger than the interatomic distance, do not obey Poisson's law but are more or less correlated.
- For penetration through crystals under channeling conditions, impact parameters are typically not selected at random; in particular, the frequency of close collisions is drastically reduced.

Two examples illustrate the first type of behavior. Consider stopping in a diatomic molecular gas and assume Bragg's additivity rule to be strictly fulfilled. Physically this implies that all changes in the electronic structure of the atoms in the molecule compared to isolated atoms can be ignored in evaluating stopping parameters, and the only molecular property left over is their mutual spatial correlation that may be assumed fixed at a distance D, while the orientation may be assumed random.

For such a system the following results were derived by Sigmund (1976) for mean energy loss and straggling,

$$\langle \Delta E \rangle = n\ell(S_1 + S_2) \tag{12.19}$$

$$\Omega^2 = n\ell \left(W_1 + W_2 + \frac{S_1 S_2}{2\pi D^2} \cdots \right), \tag{12.20}$$

if the interaction range is limited to $\ll D$, where S_1, S_2 are stopping cross sections and n the number of molecules per volume. Since W_1 and W_2 are almost constant as a function of the beam energy, the correlation term in (12.20) is most significant around the stopping maximum, at which the assumption of a short interaction range may not be far from being fulfilled.

In physical terms, the correlation in space of the target atoms implies that ion-target interactions come in pairs, which is in clear contradiction with Poisson statistics. While the average energy loss is unaffected, fluctuations become enhanced. Note that this violation of Poisson statistics is caused by adopting target *atoms* as the basic unit. The problem evaporates if the basic entity is taken to be the target *molecule*.

A molecular gas may serve as a model for a solid since interatomic distances have comparable magnitudes. For a monoatomic medium, Sigmund (1978) found

$$\langle \Delta E \rangle = n\ell S \tag{12.21}$$

$$\Omega^2 = n\ell \left(W + 2nS^2 \int_0^\infty dr\, [g(r) - 1] \cdots \right), \tag{12.22}$$

where $g(r)$ is the pair correlation of the material, normalized according to

$$n \int \mathrm{d}^3 \boldsymbol{r} \left(g(r) - 1 \right) = 1. \qquad (12.23)$$

Again the assumption enters that the interaction be smaller than the internuclear distance.

Unlike in (12.20), (12.22) yields *reduced* fluctuations because of the greater regularity of the atomic arrangement in a closely-packed solid than in a random gas.

More quantitative evaluation of correlation effects have been reported by Sigmund (1991) and, for a specific system, by Grande and Schiwietz (1991).

References

Allison, S. K. (1958). "Experimental results on charge-changing collisions of hydrogen and helium atoms and ions at kinetic energies above 0.2 keV," Rev. Mod. Phys. **30**, 1137–1168.

Bohr, N. (1948). "The penetration of atomic particles through matter," Mat. Fys. Medd. Dan. Vid. Selsk. **18 no. 8**, 1–144.

Efken, B., Hahn, D., Hilscher, D. and Wüstefeld, G. (1975). "Energy loss and energy loss straggling of N, Ne, and Ar ions in thin targets," Nucl. Instrum. Methods **129**, 219–225.

Glazov, L. and Sigmund, P. (1997). "Energy-loss spectra of charged particles in the presence of charge exchange," Nucl. Instrum. Methods B **125**, 110–115.

Glazov, L. G. (2000). "Energy-loss spectra of swift ions," Nucl. Instrum. Methods B **161**, 1–8.

Glazov, L. G. (2002). "Multiple-peak structures in energy-loss spectra of swift ions," Nucl. Instrum. Methods B **193**, 56–65.

Grande, P. L. and Schiwietz, G. (1991). "Impact-parameter dependence of electronic energy loss and straggling of incident bare ions on H and He atoms by using the coupled-channel method," Phys. Rev. A **44**, 2984–2992.

Landau, L. (1944). "On the energy loss of fast particles by ionization," J. Phys. USSR **8**, 201.

Lindhard, J. (1985). "On the theory of energy loss distributions for swift charged particles," Phys. Scr. **32**, 72–80.

Närmann, A. and Sigmund, P. (1994). "Statistics of energy loss and charge exchange of penetrating particles: Higher moments and transients," Phys. Rev. A **49**, 4709–4715.

Sigmund, P. (1976). "Energy loss and angular spread of ions traversing matter," Ann. Israel Phys. Soc. **1**, 69–120.

Sigmund, P. (1978). "Statistics of particle penetration," Mat. Fys. Medd. Dan. Vid. Selsk. **40 no. 5**, 1–36.

Sigmund, P. (1991). "Statistics of charged-particle penetration," A. Gras-Marti, H. M. Urbassek, N. Arista and F. Flores, eds., *Interaction of charged particles with solids and surfaces*, volume 271 of *NATO ASI Series*, 73–144 (Plenum Press, New York).

Sigmund, P. (1992). "Statistical theory of charged-particle stopping and straggling in the presence of charge exchange," Nucl. Instrum. Methods B **69**, 113–122.

Sigmund, P. (1994). "Analysis of charge-dependent stopping of swift ions," Phys. Rev. A **50**, 3197–3201.

Sigmund, P. and Winterbon, K. B. (1985). "Energy loss spectrum of swift charged particles penetrating a layer of material," Nucl. Instrum. Methods B **12**, 1–16.

Symon, K. (1948). *Fluctuations in energy lost by high energy charged particles in passing through matter*, Ph.D. thesis, Harvard University.

Tschalär, C. (1968). "Straggling distributions of large energy losses," Nucl. Instrum. Methods **61**, 141–156.

Vavilov, P. V. (1957). "Ionization losses of high-energy heavy particles," Zh. Eksp. Teor. Fiz. **32**, 920–923, [English translation: Sov. Phys. JETP **5**, 749-751 (1957)].

Winterbon, K. B. (1977). "Electronic energy loss and charge-state fluctuations of swift ions," Nucl. Instrum. Methods **144**, 311–315.

Winterbon, K. B., Sigmund, P. and Sanders, J. B. (1970). "Spatial distribution of energy deposited by atomic particles in elastic collisions," Mat. Fys. Medd. Dan. Vid. Selsk. **37 no. 14**, 1–73.

13 Straggling

13.1 General Survey

Statistical aspects of energy-loss straggling have been considered in Ch. 12. The present chapter is devoted to atomistic aspects, i.e., the contributions of various stopping processes to straggling and appropriate input into numerical evaluations. Attention will be paid primarily to the straggling parameter $W = \int w^2 d\sigma(w)$ for a frozen charge and the evaluation of the variance Ω^2. Results will also be reported on predicted energy-loss spectra mainly for thin targets.

It was pointed out in Chap. 2 that (2.14) is less comprehensive than (2.8). One reason for this is the correlation effect discussed in the previous section. In addition there is a significant difference between classical and quantal evaluations that is not present in case of the stopping cross section. Expressed as integrations over the impact parameter, (2.8) and (2.14) read

$$S = \int 2\pi p dp \langle w \rangle (p) \tag{13.1}$$

$$W = \int 2\pi p dp \langle w^2 \rangle (p). \tag{13.2}$$

In a classical calculation, the fluctuation in energy loss at a given impact parameter can be set equal to zero so that $\langle w^2 \rangle (p) \equiv [\langle w \rangle (p)]^2$. Hence, in the presence of significant quantal fluctuations at a given impact parameter, a classical calculation will tend to underestimate straggling.

The occurrence of the factor w^2 has the consequence of a greater significance of large energy losses in straggling than in the stopping cross section, i.e., close collisions tend to dominate (Bohr, 1915). Since the stopping cross section for unscreened Coulomb interaction diverges only logarithmically at small energy transfers, this means that the role of binding of target electrons is much less important in straggling than in stopping.

At the same time, more attention needs to be paid to nuclear energy losses. Reference is made to work by Glazov and Sigmund (2003).

13.2 Point Charge

A general reference is the result of Bohr (1915),

$$W_B = 4\pi Z_1^2 Z_2 e^4, \tag{13.3}$$

which is found for unrestricted Coulomb scattering. It is easily verified that binding gives rise to a correction term of the relative order $(Z_1 e^2 \omega / m v^3)^2$ within the Bohr theory, i.e., a term that is negligible compared to shell and Barkas-Andersen corrections.

In view of the dominance of close collisions, the shell correction may reliably be evaluated from kinetic theory (Sigmund, 1982),

$$W(v) = \left\langle \frac{3[\boldsymbol{v} \cdot (\boldsymbol{v} - \boldsymbol{v}_e)]^2 - (\boldsymbol{v} - \boldsymbol{v}_e)^2 v^2}{2v|\boldsymbol{v} - \boldsymbol{v}_e|^3} W_0(|\boldsymbol{v} - \boldsymbol{v}_e|) \right.$$
$$\left. -m \frac{[\boldsymbol{v} \cdot (\boldsymbol{v} - \boldsymbol{v}_e)]^2 - (\boldsymbol{v} - \boldsymbol{v}_e)^2 v^2}{v|\boldsymbol{v} - \boldsymbol{v}_e|} S_0(|\boldsymbol{v} - \boldsymbol{v}_e|) \right\rangle_{\boldsymbol{v}_e}, \tag{13.4}$$

where W_0 and S_0 denote the uncorrected straggling parameter and stopping cross section, respectively.

Evaluations have focused on asymptotic expansions in powers of $\langle v_e^2 \rangle / v^2$ where, according to Sigmund (1982),

$$W(v) = W_0(v) + \frac{\langle v_e^2 \rangle}{v^2} \left[-W_0(v) + \frac{1}{6} v^2 \frac{d^2}{dv^2} W_0(v) + \frac{2}{3} m v^2 S_0(v) \ldots \right]. \tag{13.5}$$

After insertion of W_B for W_0 this reduces to

$$W(v) = W_B \left[1 + \frac{\langle v_e^2 \rangle}{v^2} \left(\frac{2}{3} L_0(v) - 1 \right) \ldots \right], \tag{13.6}$$

where L_0 is the uncorrected stopping number. A correction of this type, with the factor $2/3$, was first derived by Livingston and Bethe (1937).

Extension of Bohr's estimate (Bohr, 1915) into the relativistic regime adds a factor of $\gamma^2 = 1/(1 - v^2/c^2)$ to (13.3). The Born approximation, on the other hand, leads to (Fano, 1963)

$$\Omega^2 = n\ell W_B \left[\frac{1 - v^2/2c^2}{1 - v^2/c^2} + \frac{2}{3} \frac{\langle |\sum \boldsymbol{v}_e|^2 \rangle}{Z_2 v^2} \ln \frac{2mv^2}{I_1} \right], \tag{13.7}$$

where the sum goes over all electrons of a target atom.

The factor $\langle |\sum \boldsymbol{v}_e|^2 \rangle / Z_2 v^2$ reduces to $\langle v_e^2 \rangle / v^2$ if correlations between target electrons can be neglected. Moreover,

$$\ln I_1 = \frac{\sum_j f_j \hbar\omega_j \ln(\hbar\omega_j)}{\sum_j f_j \hbar\omega_j}. \tag{13.8}$$

Equation (13.7) contains an additional relativistic correction factor and an I-value under the Bethe logarithm that is distinct for straggling. Inokuti & al. (1981) estimated

$$I_1 \simeq Z_2^{1.6384} \times 33.1 \text{ eV} \tag{13.9}$$

for $Z_2 \leq 38$.

In view of the dominance of close collisions, the Bloch correction deserves special attention. Lindhard and Sørensen (1996) showed that there is no Bloch correction to (13.7) within nonrelativistic theory. Their estimate disregarded the shell correction. Titeica (1937), on the basis of Bloch's formalism, derived an additional term of the form

$$W(v) = W_B \left[1 + \frac{\langle v_e^2 \rangle}{v^2} \left(\frac{2}{3} \left[L_0(v) \right.\right.\right.$$
$$\left.\left.\left. + \psi(1) - \text{Re} \left(1 + \text{i} \frac{Z_1 v_0}{v} \right) \right] - 1 \right) \right], \tag{13.10}$$

to replace (13.6). This expression was confirmed by Sigmund (1982) on the basis of (13.4). Thus, Titeica's term is to be regarded as a shell correction. Its existence is entirely compatible with the conclusion of Lindhard and Sørensen (1996).

A substantial correction was found by Lindhard and Sørensen (1996) for the relativistic regime. An expression was derived for the straggling parameter following the lines that led to (5.15). Figure 13.1 shows that pronounced deviations from relativistic Bohr straggling as well as from the perturbational expression may be expected for $v/c \gtrsim 0.1$ and that the sign of the deviations from the Bohr value depends on Z_1.

Nonrelativistic model calculations based on the dielectric theory (Bonderup and Hvelplund, 1971; Chu, 1976; Sigmund and Fu, 1982) and the harmonic-oscillator model (Sigmund and Haagerup, 1986) were geared toward light ions. The range of validity of the numerical results of those evaluations must be quite restricted in case of heavy ions (cf. Fig. 4.2). The same statement applies to calculations of impact-parameter-dependent straggling by Kabachnik (1993).

Estimates of the Barkas-Andersen correction to straggling have now become available. First indications of a shoulder of the Bethe-Livingston type were found by Glazov & al. (2002), and a detailed study is due to Sigmund and Schinner (2002). Figure 13.2 shows straggling calculated for protons and antiprotons as well as the average between the two. It is seen that in the absence of a shell correction, a pronounced shoulder forms for protons, while the average curve, representing protons or antiprotons in the absence of a Barkas-Andersen correction, has the form that is common from standard theory (Bonderup and Hvelplund, 1971). However, the shoulder is efficiently wiped out by the shell correction.

The influence of the Fermi density effect on straggling is expected to be insignificant because of its collective nature.

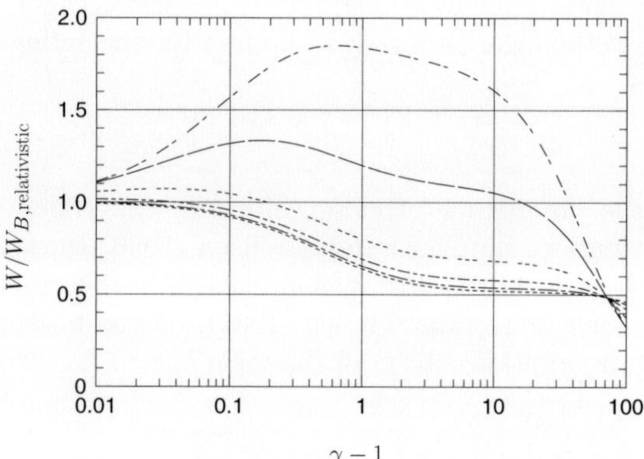

Fig. 13.1. Straggling for stripped heavy ions at relativistic velocities. Plotted is the ratio between the result of Lindhard and Sørensen (1996) and the relativistic expression by Bohr (1915), $W_{B,\text{relativistic}} = 4\pi Z_1^2 Z_2 e^4 \gamma^2$ for U, Gd, Ge, Si, O and Be (top to bottom).

13.3 Dressed Ions

For dressed ions a contribution from charge-exchange straggling needs to be added to 'collisional straggling', i.e., due to excitation of target atoms (Flamm and Schumann, 1916). A comprehensive statistical scheme (Sigmund, 1992) to treat these processes – also incorporating projectile excitation/deexcitation – was mentioned in Sect. 12.3.

Qualitative trends on collisional straggling for dressed ions may be extracted from studies by Kaneko (1990) (U in C and O in Al), Yang (1994) (C in C) and Glazov & al. (2002) (O in C). The models differ in detail but none of them makes full use of the theoretical schemes discussed in Sect. 9.

A more systematic theoretical study has been performed by Sigmund and Schinner (2002) on the basis of the binary theory.

Figure 13.3 shows relative straggling for frozen-charge Li ions in carbon. It is seen that screening affects the shape of the straggling parameter below the shoulder region, both for the shoulder generated by the Barkas-Andersen effect in the left graph and by the shell correction in the right one. The actual degree of screening appears to have surprisingly little influence. Figure 13.4 shows a more gradual variation for a heavier ion (Ar in C).

The effect of intra-atomic correlation on straggling was first studied by Besenbacher & al. (1980) for He ions. This effect goes roughly with the square of the stopping force. Preliminary estimates by Sigmund and Schinner (2002) indicate a maximum correction by correlation of $\sim 40\,\%$ of the Bohr value for He-Si and less for helium in lighter materials. These values are somewhat

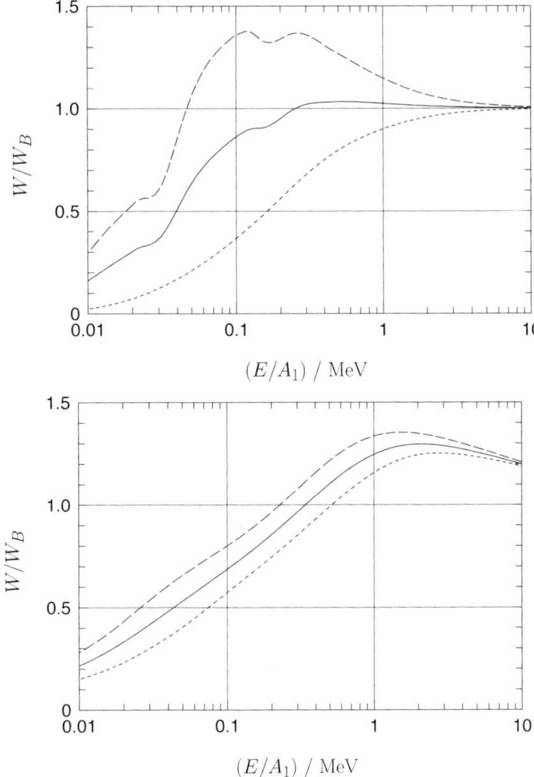

Fig. 13.2. Relative straggling W/W_B for protons (dashed lines) and antiprotons (dotted lines) in silicon, and average (solid lines). Without (top) and with (bottom) shell correction. From Sigmund and Schinner (2002)

higher than those of Besenbacher & al. (1980) based on a free-electron gas model. Corresponding values for argon ions were found to be about a factor of two higher. Experimental data that could confirm or reject such high corrections for correlation are desirable.

A systematic effort has been made to determine the charge-exchange contribution to straggling. Estimates have been based either on solutions of the transport equation for energy loss in the presence of charge exchange presented by Winterbon (1977) or on Monte Carlo simulations involving cross sections for capture and loss. While numerous studies were devoted to helium ions, heavier ions were addressed by Vollmer (1974); Efken & al. (1975); Cowern & al. (1979); Kaneko (1988); Yang and MacDonald (1993) in the analysis of experiments. Figure 13.5 from Cowern & al. (1979) shows

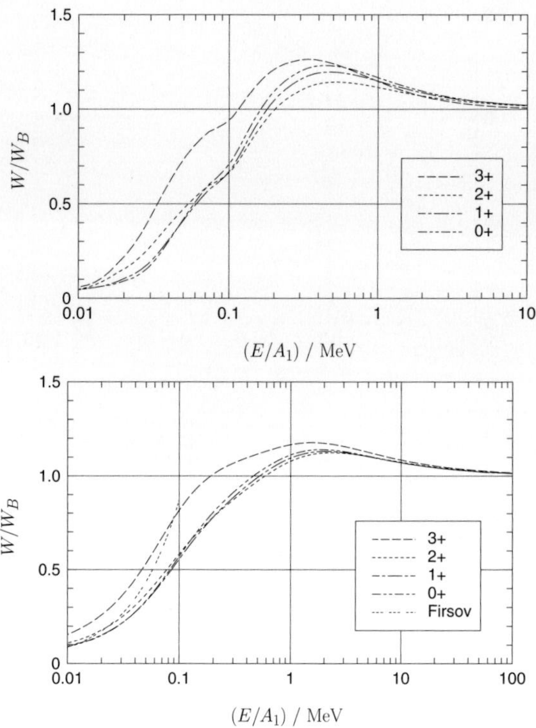

Fig. 13.3. Relative straggling for lithium in carbon for frozen charge states 3+ to 0. Without (left) and with (right) shell correction. From Sigmund and Schinner (2002)

the case of carbon on aluminium, for which collisional and charge-exchange straggling are comparable in magnitude.

Ogawa & al. (1992, 1993, 1996) measured energy-loss spectra for carbon, oxygen and lithium ions, respectively, penetrating through thin carbon foils at high speed in which the majority of the ions is fully-stripped in charge equilibrium. Especially the lithium experiments at 10 MeV/u attracted theoretical attention. Foil thicknesses can be varied in these experiments such as to identify individual capture-loss cycles in the measured spectra. The shape of those individual peaks must then be governed by collisional straggling. In accordance with an earlier analysis (Glazov and Sigmund, 1997) of measurements on helium (Ogawa & al., 1991), it was found that experimental energy resolution did not allow information to be extracted about collisional straggling. Claims to the contrary by Balashov & al. (1997) were demonstrated to be in error (Glazov and Sigmund, 2000).

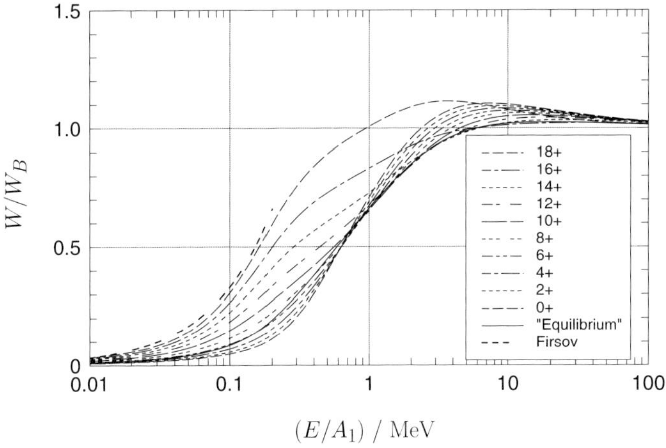

(E/A_1) / MeV

Fig. 13.4. Same as Fig. 13.3 for argon in carbon. From Sigmund and Schinner (2002)

13.4 Low-Speed Ions

The theoretical schemes discussed in Sect. 8 could in principle be applied to straggling, but only few studies have been made. In particular, the theory underlying formula (8.4) by Lindhard and Scharff (1961) has never been applied to straggling.

Integration of (8.2) from the theory of Firsov (1959) yields

$$W = \int 2\pi p dp w^2(v, p) = 0.133\pi\hbar^2 v^2 \left(Z_1 + Z_2\right)^{8/3}. \tag{13.11}$$

Hvelplund (1971), measuring straggling in gases, found a rough overall agreement with this prediction as indicated in Fig. 13.6, but significant deviations are observed both from the predicted v^2-dependence and in absolute magnitude. However, near-perfect agreement was found for Li-He from 100 to 500 keV.

Kaneko (1990) derived an expression for straggling at low speed from the dielectric theory combined with a model for screening in an electron gas,

$$W = 12\pi v^2 \hbar^2 Z_1^{4/3} z_{\text{free}} A^2, \tag{13.12}$$

where z_{free} is the number of conduction electrons per target atom,

$$A^2 = \frac{0.56}{1 - 0.51/Z_1^{2/3} r_s}, \tag{13.13}$$

and r_s the Wigner-Seitz radius. Figure 13.7 shows a comparison with the data of Hvelplund (1971) for neon targets, but now at all velocities. Discrepancies up to more than a factor of 2 are found.

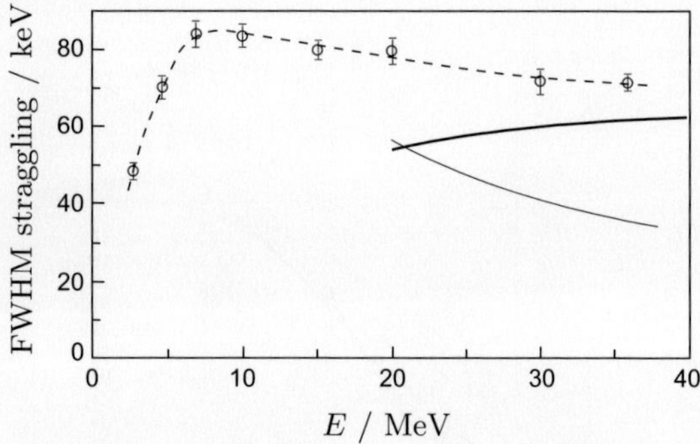

Fig. 13.5. Energy-loss straggling for ^{12}C in 217 µg/cm^2 Al. Experimental data and theoretical estimates. Upper solid line: Collisional straggling from Bethe-Livingston formula (estimated from variance). Lower solid line: Contribution from charge exchange, estimated by use of cross sections extracted from the same series of experiments. Dashed line: Total straggling $\Omega_{\mathrm{tot}} = \sqrt{\Omega_{\mathrm{coll}}^2 + \Omega_{\mathrm{chex}}^2}$. From Cowern & al. (1979).

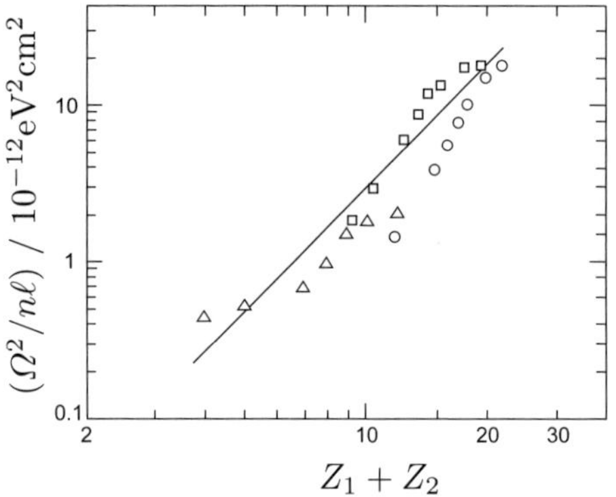

Fig. 13.6. Energy-loss straggling in He (triangles), air (squares) and neon (circles) at $v = 0.9\, v_0$ compared with (13.11) according to Hvelplund (1971).

Figure 13.8 shows calculations on the basis of the binary theory by Sigmund and Schinner (2002) compared with measurements for lithium ions in He, Ne and Ar by Andersen & al. (1978). Tolerable agreement is found for Li-Ar and Li-Ne above v_0, while the theory appears to overestimate straggling for the Li-He system.

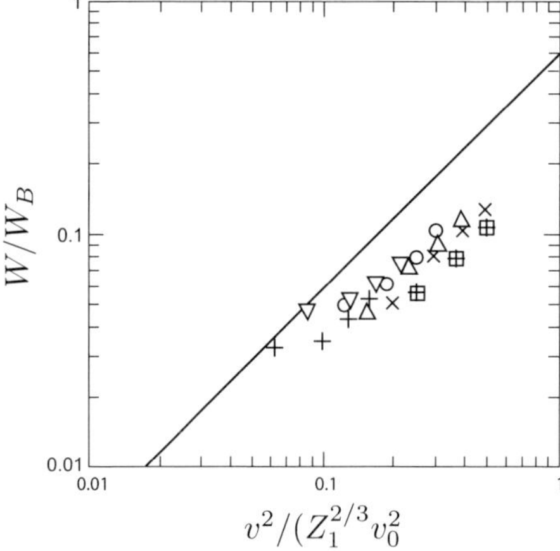

Fig. 13.7. Comparison of straggling data for B ($\bigcirc$), C ($\times$), N ($\triangle$), O ($\square$), Ne ($\triangledown$) and Mg ($+$) ions in neon with (13.13). From Kaneko (1990).

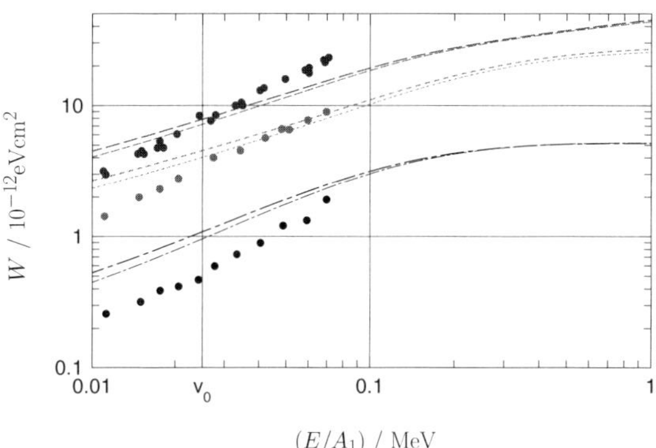

Fig. 13.8. Straggling for Li ions in noble gases: Calculations by binary theory disregarding correlation effect compared with measurements of Andersen & al. (1978) on Ar, Ne and He (top to bottom). Thick lines: ion in mean equilibrium charge. Thin lines: neutral ion. From Sigmund and Schinner (2002)

References

Andersen, H. H., Besenbacher, F. and Knudsen, H. (1978). "Stopping power and straggling of 65-500 keV Lithium ions in H_2, He, CO_2, N_2, O_2, Ne, Ar, Kr, and Xe," Nucl. Instrum. Methods **149**, 121–127.

Balashov, V. V., Bibikov, A. V. and Bodrenko, I. V. (1997). "Effect of charge exchange on the energy distribution of fast multiply charged ions propagating through matter," Zh. Eksp. Teor. Fiz. **111**, 2226–2236, [English translation: JETP **84**, 1215-1220 (1997)].

Besenbacher, F., Andersen, J. U. and Bonderup, E. (1980). "Straggling in energy loss of energetic hydrogen and helium ions," Nucl. Instrum. Methods **168**, 1–15.

Bohr, N. (1915). "On the decrease of velocity of swiftly moving electrified particles in passing through matter." Philos. Mag. **30**, 581–612.

Bonderup, E. and Hvelplund, P. (1971). "Stopping power and energy straggling for swift protons," Phys. Rev. A **4**, 562–589.

Chu, W. K. (1976). "Calculation of energy straggling for protons and helium ions," Phys. Rev. A **13**, 2057–2060.

Cowern, N. E. B., Sofield, C. J., Freeman, J. M. and Mason, J. P. (1979). "Energy straggling of $3 - 36 - $ MeV ^{12}C ions in aluminum," Phys. Rev. A **19**, 111–115.

Efken, B., Hahn, D., Hilscher, D. and Wüstefeld, G. (1975). "Energy loss and energy loss straggling of N, Ne, and Ar ions in thin targets," Nucl. Instrum. Methods **129**, 219–225.

Fano, U. (1963). "Penetration of protons, alpha particles, and mesons," Ann. Rev. Nucl. Sci. **13**, 1–66.

Firsov, O. B. (1959). "A qualitative interpretation of the mean electron excitation energy in atomic collsions," Zh. Eksp. Teor. Fiz. **36**, 1517–1523, [English translation: Sov. Phys. JETP **9**, 1076-1080 (1959)].

Flamm, L. and Schumann, R. (1916). "Die Geschwindigkeitsabnahme der α-Strahlen in Materie." Ann. Physik **50**, 655.

Glazov, L. and Sigmund, P. (1997). "Energy-loss spectra of charged particles in the presence of charge exchange," Nucl. Instrum. Methods B **125**, 110–115.

Glazov, L. and Sigmund, P. (2000). "Energy-loss spectra of charged particles in the presence of charge exchange: addendum on ^{6}Li spectra," Nucl. Instrum. Methods B **170**, 39–44.

Glazov, L. G. and Sigmund, P. (2003). "Nuclear stopping in transmission measurements," Nucl. Instrum. Methods B **207**, 240–256.

Glazov, L. G., Sigmund, P. and Schinner, A. (2002). "Statistics of heavy-ion stopping," Nucl. Instrum. Methods B **195**, 183–187.

Hvelplund, P. (1971). "Energy loss and straggling of 100-500 keV atoms with $2 \leq Z_1 \leq 12$ in various gases," Mat. Fys. Medd. Dan. Vid. Selsk. **38 no. 4**, 1–25.

Inokuti, M., Dehmer, J. L., T., B. and Hanson, J. D. (1981). "Oscillator-strength moments, stopping powers, and total inelastic- scattering cross sections of all atoms through strontium," Phys. Rev. A **23**, 95–109.

Kabachnik, N. M. (1993). "Screening and antiscreening in the semiclassical description of ionization in fast ion-atom collisions," J. Phys. B **26**, 3803–3814.

Kaneko, T. (1988). "Energy loss and straggling accompanied by charge exchange," Nucl. Instrum. Methods B **33**, 151–154.

Kaneko, T. (1990). "Energy loss and straggling of heavy ions in matter," Nucl. Instrum. Methods B **48**, 83–86.

Lindhard, J. and Scharff, M. (1961). "Energy dissipation by ions in the keV region," Phys. Rev. **124**, 128–130.

Lindhard, J. and Sørensen, A. H. (1996). "On the relativistic theory of stopping of heavy ions," Phys. Rev. A **53**, 2443–2456.

Livingston, M. S. and Bethe, H. A. (1937). "Nuclear physics. C. Nuclear dynamics, experimental," Rev. Mod. Phys. **9**, 245–390.

Ogawa, H., Katayama, I., Ikegami, H., Haruyama, Y., Aoki, A., Tosaki, M., Fukuzawa, F., Yoshida, K., Sugai, I. and Kaneko, T. (1991). "Direct Measurement of fixed-charge stopping power for 32-MeV He^{3+} in a charge-state nonequilibrium region," Phys. Rev. B **43**, 11370–11376.

Ogawa, H., Katayama, I., Sugai, I., Haruyama, Y., Saito, M., Yoshida, K. and Tosaki, M. (1996). "Energy loss of high velocity Li-6^{2+} ions in carbon foils in charge state non-equilibrium region," Nucl. Instrum. Methods B **115**, 66–69.

Ogawa, H., Katayama, I., Sugai, I., Haruyama, Y., Saito, M., Yoshida, K., Tosaki, M. and Ikegami, H. (1993). "Charge state dependent energy loss of high velocity oxygen ions in the charge state non-equilibrium region," Nucl. Instrum. Methods B **82**, 80–84.

Ogawa, H., Katayama, I., Sugai, I., Haruyama, Y., Tosaki, M., Aoki, A., Yoshida, K. and Ikegami, H. (1992). "Charge state dependent energy loss of high velocity carbon ions in the charge state non-equilibrium region," Phys. Lett. A **167**, 487–492.

Sigmund, P. (1982). "Kinetic theory of particle stopping in a medium with internal motion," Phys. Rev. A **26**, 2497–2517.

Sigmund, P. (1992). "Statistical theory of charged-particle stopping and straggling in the presence of charge exchange," Nucl. Instrum. Methods B **69**, 113–122.

Sigmund, P. and Fu, D.-J. (1982). "Energy loss straggling of a point charge penetrating a free-electron gas," Phys. Rev. A **25**, 1450.

Sigmund, P. and Haagerup, U. (1986). "Bethe stopping theory for a harmonic oscillator and Bohr's oscillator model of atomic stopping," Phys. Rev. A **34**, 892–910.

Sigmund, P. and Schinner, A. (2002). "Barkas effect, shell correction, screening and correlation in collisional energy-loss straggling of an ion beam," Europ. Phys. J. D 201–209.

Titeica, S. (1937). "Sur Les Fluctuations de Parcours Des Rayons Corpusculaires," Bull. Soc. Roumaine Phys. **38**, 81.

Vollmer, O. (1974). "Der Einfluss der Ladungsfluktuationen auf die Energieverlustverteilung geladener Teilchen," Nucl. Instrum. Methods **121**, 373–377.

Winterbon, K. B. (1977). "Electronic energy loss and charge-state fluctuations of swift ions," Nucl. Instrum. Methods **144**, 311–315.

Yang, Q. (1994). "Partial stopping power and straggling effective charges of heavy ions in condensed matter," Phys. Rev. A **49**, 1089–1095.

Yang, Q. and MacDonald, R. J. (1993). "Energy loss and straggling of heavy ions in condensed matter," Nucl. Instrum. Methods B **83**, 303–310.

14 Multiple Scattering

The statistical scheme outlined in Chap. 12 also applies to multiple angular deflection, or *multiple scattering*, except that one deals with a two-dimensional problem. Moreover, energy loss is a one-way process for ions slowing down in cold matter, but that aspect is of minor significance in the general formalism. Multiple scattering was first studied along these lines by Bothe (1921). The theory was developed subsequently by Molière (1948), Bethe (1953), Meyer (1971), Sigmund and Winterbon (1974) and Amsel & al. (2003).

Consider a monochromatic, well collimated beam penetrating through a foil of uniform thickness x that is small enough so that variations of the scattering cross section because of decreasing beam energy can be neglected. Individual scattering events are characterized by a differential scattering cross section $d\sigma(\phi) = K(\phi)d^2\phi$, where ϕ is a scattering angle in the laboratory frame of reference.

For heavy ions, multiple-scattering distributions are typically of interest for angular cones with opening angles of a few degrees. Hence the small-angle approximation, $\sin\phi \simeq \tan\phi \simeq \phi$, will be justified for many purposes. This approximation is implied here except where stated otherwise.

Let $F(\alpha, x)2\pi\alpha d\alpha$ be the probability per beam particle to be scattered into a solid angle $(\alpha, d\alpha)$ after penetrating a foil thickness x. Then the Bothe-Landau formula (12.1) may be rewritten in terms of scattering angles and reads (Bothe, 1921; Sigmund and Winterbon, 1974)

$$F(\alpha, x) = \frac{1}{2\pi} \int_0^\infty k dk\, J_0(k\alpha) e^{-nx\sigma(k)}, \tag{14.1}$$

where n denotes the density of scattering centers, J_0 a Bessel function in standard notation, and

$$\sigma(k) = \int_0^\infty d\sigma(\phi)\left[1 - J_0(k\phi)\right]. \tag{14.2}$$

Alternatively one may write (Goudsmit and Saunderson, 1940)

$$F(\alpha, \ell) = \frac{1}{4\pi} \sum_{\mu=0}^\infty (2\mu + 1) P_\mu(\cos\alpha) e^{-n\ell\sigma_\mu}, \tag{14.3}$$

where $P_l(\cos\phi)$ are Legendre polynomials,

$$\sigma_\mu = \int d\sigma(\phi)\left[1 - P_\mu(\cos\phi)\right],\tag{14.4}$$

and ℓ denotes the travelled pathlength which, within the small-angle approximation, is equal to the foil thickness[1].

For heavy ions, multiple angular deflection is due to nuclear collisions with the exception of the case of channeling in which nuclear interactions are strongly suppressed.

Computation of multiple-scattering profiles is simplified substantially by the use of scaling laws of the type discussed for nuclear stopping in section 4.8. Convenient variables for scattering and target thickness are (Molière, 1948; Meyer, 1971; Sigmund and Winterbon, 1974)

$$\tilde{\alpha} = \frac{Ea}{2Z_1Z_2e^2}\,\alpha;\quad \tau = \pi a^2 nx,\tag{14.5}$$

where a is the screening radius of the interatomic potential. Within the small-angle approximation the multiple-scattering distribution as expressed by (14.1) reduces to a universal function of $\tilde{\alpha}$ and τ that is determined by the adopted interatomic potential. This function has been tabulated for $0.001 \leq \tau \leq 2000$ by Sigmund and Winterbon (1974)[2].

Within the small-angle approximation, the multiple-scattering half-width likewise obeys a scaling law, i.e.,

$$\tilde{\alpha}_{1/2} = g(\tau),\tag{14.6}$$

where g is a universal function determined by the interatomic potential. This relation is shown in Fig. 14.1 for Lenz-Jensen and Thomas-Fermi interaction specified by Table 10.2. Also included are results for power-law scattering governed by (10.14).

Comments made on page 101 in connection with energy-loss straggling on the transition from thick to very-thick targets, in particular (12.7), also apply to small-angle multiple scattering. The point has been studied in more detail by Valdes and Arista (1994).

In addition to angular deflection, also the lateral spread of an initially narrow beam can be of interest. Lateral and angular distributions are closely related and can be mapped upon each other. This mapping invokes power-law scattering with $m = m(\tau)$. For details the reader is referred to Marwick and Sigmund (1975).

---

[1]Unlike (14.1) which assumes small angles α, (14.3) holds for all angles in principle. However, the pathlength itself is not necessarily measurable. Therefore, also (14.3) may be limited to small angles where the pathlength ℓ is close to the foil thickness.

[2]with an important erratum for $\tau = 0.05$ and 0.10 (Sigmund and Winterbon, 1975).

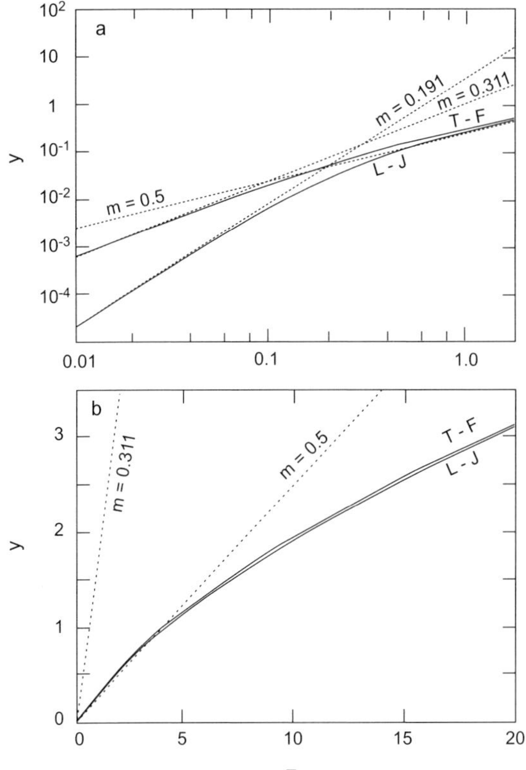

Fig. 14.1. Multiple-scattering halfwidth *versus* thickness in dimensionless units according to Sigmund and Winterbon (1974). Thomas-Fermi and Lenz-Jensen interaction and power laws. The two graphs cover different thickness ranges.

A code calculating multiple-scattering angular profiles according to Sigmund and Winterbon (1974) has become available (Eyeberger, 1999). A more extensive code, also providing lateral distributions and profiles projected on a plane and based on slightly different physical input is available from Amsel *&* al. (2003).

The prime modification of the above scheme in the relativistic regime is the replacement of the kinetic energy E in (14.5) by (Bohr, 1948)

$$E \to \frac{1}{2}\gamma M_1 v^2; \quad \gamma = \frac{1}{\sqrt{1 - v^2/c^2}}. \tag{14.7}$$

In addition, screening might not be described adequately by Thomas-Fermi interaction (Schwab *&* al., 1990), and deviations from Coulomb scattering due to the finite size of the nucleus can become relevant (Williams, 1939; Bohr, 1948).

References

Amsel, G., Battistig, G. and L'Hoir, A. (2003). "Small angle multiple scattering of fast ions, physics, stochastic theory and numerical calculations," Nucl. Instrum. Methods B **201**, 325–388, URL www.mfm.kfki.hu/ms.

Bethe, H. A. (1953). "Moliere's theory of multiple scattering," Phys. Rev. **89**, 1256–1266.

Bohr, N. (1948). "The penetration of atomic particles through matter," Mat. Fys. Medd. Dan. Vid. Selsk. **18 no. 8**, 1–144.

Bothe, W. (1921). "Theorie der Zerstreuung der α-Strahlen über kleine Winkel." Z. Phys. **4**, 300.

Eyeberger, L. (1999). "SWIMS, Sigmund and Winterbon multiple scattering of ion beams," URL www.nea.fr/abs/html/ests0682.html.

Goudsmit, S. and Saunderson, J. L. (1940). "Multiple scattering of electrons," Phys. Rev. **57**, 24–29.

Marwick, A. D. and Sigmund, P. (1975). "Small-angle multiple scattering of ions in the screened Coulomb region. 2. Lateral spread," Nucl. Instrum. Methods **126**, 317–323.

Meyer, L. (1971). "Plural and multiple scattering of low-energy heavy particles in solids," phys. stat. sol. (b) **44**, 253–268.

Molière, G. (1948). "Theorie der Streuung schneller geladener Teilchen. II. Mehrfach- und Vielfachstreuung," Z. Naturforsch. **3a**, 78–97.

Schwab, T., Geissel, H., Armbruster, P., Gillibert, A., Mittig, W., Olson, R. E., Winterbon, K. B., Wollnik, H. and Munzenberg, G. (1990). "Energy and angular distributions for Ar-ions penetrating solids," Nucl. Instrum. Methods B **48**, 69–74.

Sigmund, P. and Winterbon, K. B. (1974). "Small-angle multiple scattering of ions in the screened Coulomb region, I. Angular distributions," Nucl. Instrum. Methods **119**, 541–557.

Sigmund, P. and Winterbon, K. B. (1975). "Erratum," Nucl. Instrum. Methods **125**, 491.

Valdes, J. E. and Arista, N. R. (1994). "Energy-loss effects in multiple-scattering distributions of ions in matter," Phys. Rev. A **49**, 2690–2696.

Williams, E. J. (1939). "Concerning the scattering of fast electrons and of cosmic-ray particles," Proc. Roy. Soc. A **169**, 531–572.

15 Restricted Nuclear Stopping

This chapter addresses the central question of experimentally separating nuclear from electronic stopping. Reference is made specifically to transmission experiments in which the energy-loss spectrum of ions emerging downstream from the target is measured over a narrow angular interval around the initial beam direction. Ions that have undergone a violent nuclear collision will typically be deflected outside the acceptance angle of the detector. Therefore the full (unrestricted) nuclear stopping cross section will not contribute to the observed energy-loss spectrum. Conversely, ions exiting in the forward direction will typically have undergone a series of small-angle nuclear-scattering events. Hence some nuclear stopping will be recorded. The corresponding *restricted nuclear-stopping cross section* can be expected to vary slowly with angle within the multiple-scattering cone. More important, it will depend sensitively on the penetration depth.

Above the shell-correction limit, where the majority of the target electrons contributes to electronic stopping, nuclear stopping accounts for less than 0.1 % of the total stopping. In fact, Fig. 4.2 indicates that nuclear stopping becomes a serious problem mainly for slow ions. Therefore, pertinent theoretical schemes to correct for nuclear stopping refer to the velocity regime in which nuclear and electronic stopping have comparable magnitudes. This is the regime in which Z_1 and Z_2 structure are observed and where noticeable phase effects and deviations from Bragg additivity can be expected.

Following Ormrod and Duckworth (1963), it has been customary in this range to measure peak rather than mean energy losses. This is motivated by the fact that observed energy-loss spectra are skewed with a pronounced high-loss tail that was asserted to be caused by nuclear stopping. Operating with the peak energy loss had been expected to provide more reliable stopping forces due to a smaller nuclear-stopping correction. On the other hand, this generates uncertainties due to the mean-to-peak ratio in pure electronic stopping, which is largely unknown in the energy range in question.

Existing theoretical or numerical schemes (Ormrod and Duckworth, 1963; Fastrup & al., 1966; Skoog, 1975; Geissel & al., 1984; Krist & al., 1984; Lennard & al., 1986; Glazov and Sigmund, 2003) address the behavior of the peak energy loss as a function of target thickness and emission angle. The standard scheme used by most experimental groups was developed by

Fastrup & al. (1966) on the basis of the Bohr-Williams theory of energy straggling and multiple scattering (Bohr, 1948; Williams, 1939, 1940). The scheme is intuitive rather than quantitative but has proven very efficient and surprisingly accurate when compared with a more rigorous scheme based on the Bothe-Landau theory, which treats both peak and mean energy losses.

15.1 Bohr-Williams Theory

The Bohr-Williams theory divides trajectories into one group made up by ions that have undergone a single violent collision at a scattering angle ϕ exceeding some limiting angle ϕ_1, and another one made up by ions that have undergone a large number of small-angle events $\phi < \phi_1$. The limiting angle is defined by

$$\phi_1^2 = nx \int_{\phi < \phi_1} \phi^2 \mathrm{d}\sigma(\phi). \tag{15.1}$$

For elastic collisions, the relation between energy loss and scattering angle is unique if the projectile mass M_1 is smaller than the target mass M_2, i.e.,

$$w = \frac{M_1}{M_2} E \phi^2 \tag{15.2}$$

at small angles. Thus, the energy-loss spectrum recorded at angles $\phi > \phi_1$ will be sharply peaked around the value given by (15.2).

Conversely, the energy-loss spectrum recorded within the cone $\phi < \phi_1$ will vary slowly with ϕ and can be approximated by a gaussian centered around

$$w_1 = \frac{M_1}{M_2} E \phi_1^2. \tag{15.3}$$

Rewriting (15.1) in terms of the energy-loss cross section $\mathrm{d}\sigma(w))$ one finds

$$w_1 = nx \int_0^{w_1} w \mathrm{d}\sigma(w). \tag{15.4}$$

Dependent on foil thickness, w_1 may be significantly smaller than the maximum recoil loss[1] $w_{\mathrm{max}} = \gamma E$. Hence, the mean energy loss observed in the forward direction will appear reduced.

The width w_1^* of the peak in the nuclear-energy-loss spectrum integrated over all angles was estimated from

$$w_1^{*\,2} = Nx \int_0^{w_1^*} w^2 \mathrm{d}\sigma(w). \tag{15.5}$$

---

[1]Concerning γ cf. footnote 2 on page 7

It was argued that energy losses exceeding w_1^* contribute to the tail of the energy-loss spectrum, while a gaussian peak is generated by collisions with $w < w_1^*$. These collisions give rise to a total energy loss

$$\Delta E_{\mathrm{nucl}} = Nx \int_0^{w_1^*} w \mathrm{d}\sigma(w), \qquad (15.6)$$

which may be identified with the peak of the energy-loss spectrum. It was then demonstrated that $w_1^* < w_1$, i.e., collisions that contribute to the peak energy loss do not give rise to loss of beam particles out of the multiple-scattering cone. Hence, it was concluded that the peak energy loss at zero angle can be identified with the peak energy loss of the angular-integrated spectrum.

This assumes negligible skewness in the electronic-stopping itself.

15.2 Bothe-Landau Theory

Glazov and Sigmund (2003) describe the joint distribution in energy loss and deflection angle in terms of a Bothe-Landau-type formula

$$F(\Delta \boldsymbol{v}, t)\mathrm{d}^3\Delta\boldsymbol{v} = \frac{\mathrm{d}^3\Delta\boldsymbol{v}}{(2\pi)^3} \int \mathrm{d}^3\boldsymbol{k}\, \mathrm{e}^{\mathrm{i}\boldsymbol{k}\cdot\Delta\boldsymbol{v}-nvt\sigma(\boldsymbol{k})}, \qquad (15.7)$$

where $\Delta \boldsymbol{v}$ is the accumulated velocity change undergone by the projectile while moving at a speed v through the medium over a time interval t, where

$$\sigma(\boldsymbol{k}) = \int \mathrm{d}\sigma(\Delta\boldsymbol{u}) \left(1 - \mathrm{e}^{-\mathrm{i}\boldsymbol{k}\cdot\Delta\boldsymbol{u}}\right) \qquad (15.8)$$

is a transport cross section and $\mathrm{d}\sigma(\Delta\boldsymbol{u})$ the differential cross section for an individual collision resulting in a velocity change $(\Delta\boldsymbol{u}, \mathrm{d}^3\Delta\boldsymbol{u})$. This expression is rigorous for a random scattering in a thin or moderately thick target, i.e., long as the energy loss is small enough so that the variation with energy of the cross section is negligible.

In terms of total energy loss ΔE and overall deflection angle α, (15.7), reads

$$F(\Delta E, \alpha, \ell) = \frac{1}{8\pi^2} \sum_{\ell=0}^{\infty} (2\mu + 1)P_\mu(\cos\theta) \times \int_{-\infty}^{\infty} \mathrm{d}k \mathrm{e}^{\mathrm{i}k\Delta E - n\ell\sigma_\mu(k)}, \qquad (15.9)$$

where $\ell = vt$ is the travelled pathlength, $F(\Delta E, \alpha, \ell)\,\mathrm{d}(\Delta E)\,2\pi\sin\alpha\,\mathrm{d}\alpha$ the joint distribution in energy loss and deflection angle normalized to 1, P_μ a Legendre polynomial and

$$\sigma_\mu(k) = \int \mathrm{d}\sigma(w) \left[1 - P_\mu(\cos\phi)\mathrm{e}^{-\mathrm{i}kw}\right]. \qquad (15.10)$$

Within the small-angle approximation one may alternatively write

$$F(\Delta E, \alpha, x) = \frac{1}{4\pi^2} \int_{-\infty}^{\infty} e^{i\varkappa \Delta E}\, \mathrm{d}\varkappa \int_0^{\infty} k\, \mathrm{d}k\, J_0(k\alpha) e^{-nx\sigma_B(k,\varkappa)} \tag{15.11}$$

and

$$\sigma_B(k,\varkappa) = \int \mathrm{d}\sigma(w) \times \left[1 - J_0(k\phi(w))e^{-i\varkappa w}\right]. \tag{15.12}$$

The two descriptions (15.9) and (15.11) are equivalent at small angles and energy losses but have different merits. Equation (15.9) is formally valid at all deflection angles, although a practical upper limit is set by the fact that the pathlength ℓ is not measurable in general. Equation (15.11) is more useful at small pathlengths where the μ–series in (15.9) converges slowly. Equation (15.9) is computationally more economical in the evaluation of moments in terms of both angle and energy, again with the exception of small pathlength. Equation (15.11), on the other hand, allows analytic estimates and has favorable scaling properties.

Table 15.1. Scaling relations obeyed by peak energy loss (first row), effective stopping cross section (second row) and stopping cross section (third row). Functions carrying a tilde stand for functions determined uniquely by the interatomic potential within Lindhard scaling.

Angle-dependent	Unrestricted
$\Delta\epsilon_p = \dfrac{\gamma}{\epsilon}\tilde{g}(\tau,\tilde{\alpha})$	$\Delta\epsilon_p = \dfrac{\gamma}{\epsilon}\tilde{g}(\tau)$
$\dfrac{\Delta\epsilon_p}{\rho} = \dfrac{1}{\epsilon}\tilde{f}(\tau,\tilde{\alpha})$	$\dfrac{\Delta\epsilon_p}{\rho} = \dfrac{1}{\epsilon}\tilde{f}(\tau)$
$\dfrac{\langle\Delta\epsilon\rangle}{\rho} = \tilde{h}(\tau,\tilde{\alpha}),\ \tau \to 0$	$\dfrac{\langle\Delta\epsilon\rangle}{\rho} = \tilde{h}(\epsilon)$

15.3 Scaling Laws

Table 15.1 shows that scaling relations obeyed by peak and mean energy losses invoke dimensionless Lindhard units for energy loss ϵ and ρ as well as for multiple scattering τ and α. It is worth noting that neither the peak

and the mean energy loss at zero scattering angle nor the angle-integrated
peak energy loss show the simple proportionality with target thickness that
is found for the angle-integrated mean energy loss.

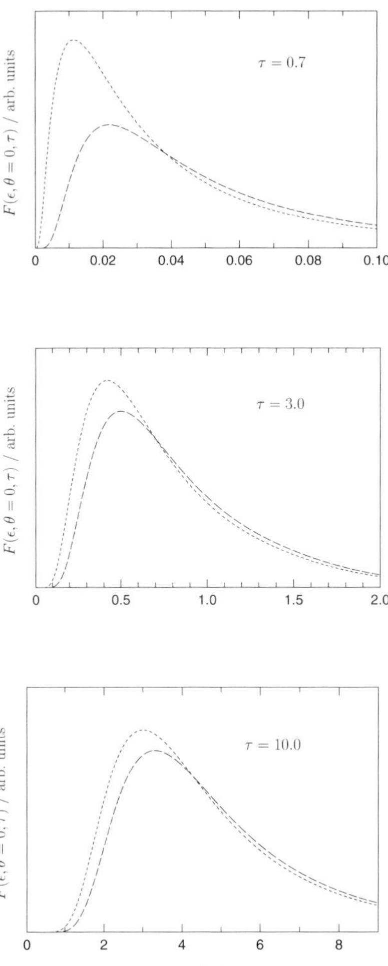

Fig. 15.1. Energy-loss spectra due to nuclear stopping at zero angle to incident-beam direction in dimensionless variables. Three values of dimensionless thickness τ. For Thomas-Fermi (dashed lines) and Lenz-Jensen (dotted lines) interaction. From Glazov and Sigmund (2003)

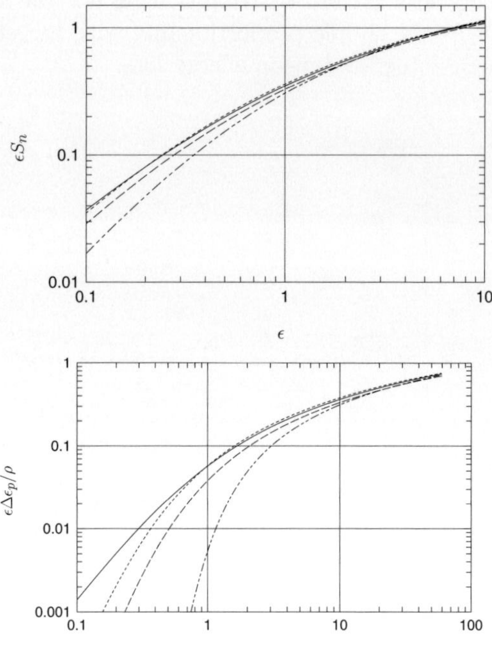

Fig. 15.2. Upper graph: Stopping cross section in dimensionless Thomas-Fermi units multiplied by ϵ, versus dimensionless energy ϵ. Lower graph: Same quantity except for replacement of mean by peak energy loss, and τ replacing ϵ as the abscissa variable. Thomas-Fermi (solid lines), Molière (dotted lines), Lenz-Jensen (dashed lines) and Bohr (dot-dashed lines) screening. From Glazov and Sigmund (2003).

15.4 Predictions

Figure 15.1 shows energy-loss spectra evaluated for Thomas-Fermi and Lenz-Jensen interactions for three representative values of the dimensionless thickness variable τ. All spectra are skewed, although skewness decreases with increasing thickness. The sensitivity to the atomic interaction potential is considerable, especially around and below the peak.

The lower part of Fig. 15.2 shows the peak energy loss in Thomas-Fermi units divided by the multiple-scattering target thickness τ. This quantity can be compared with the stopping cross section multiplied by ϵ. That is

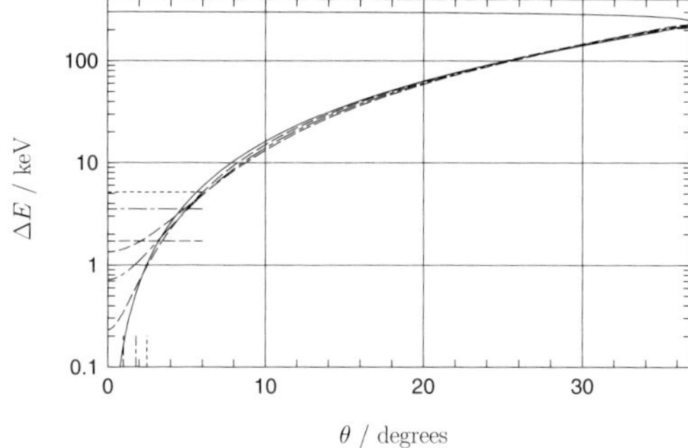

Fig. 15.3. Comparison of restricted with unrestricted nuclear energy loss for 0.8 v_0 Ne in C for foils of thickness 4.5, 9.0 and 13.5 µg/cm^2 (dashed, dot-dashed and dotted lines, respectively). Lenz-Jensen interaction assumed. Horizontal lines: Unrestricted nuclear energy loss. Broken curved lines: Calculated mean nuclear energy loss *versus* deflection angle. Thin solid line: Single scattering, upper and lower branch. Vertical lines: Multiple-scattering halfwidths. Extracted from Glazov and Sigmund (2003).

shown in the upper graph of Fig. 15.2. Although the different abscissa scales do not allow a meaningful quantitative comparison, it is evident that the peak energy loss is much more sensitive to the interaction potential than the angle-integrated mean energy loss. This is due to the lack of a contribution by violent collision events to the peak energy loss.

Figure 15.3 shows the calculated mean energy loss for three values of the foil thickness compared to the single-scattering energy transfer. As expected, the single-scattering prediction underestimates the energy loss at small angles of emergence. This is compensated at larger angles. Inspection of the multiple-scattering half-widths $\alpha_{1/2}$ – which have also been included in the graph – reveals that the point of crossover lies at $\sim 2\alpha_{1/2}$. The variation of the calculated mean energy loss within $\alpha_{1/2}$ is considerable albeit less than a factor of 2. In relative terms, the unrestricted nuclear energy loss is reduced by factors 0.150 (0.134), 0.226 (0.211), and 0.277 (0.262) for the three target thicknesses, with Lenz-Jensen values in brackets.

Figure 15.4 shows the variation with target thickness of mean energy loss and straggling at zero emergence angle. These quantities decrease faster with decreasing target thickness than the corresponding unrestricted (angle-integrated) quantities that also have been included in the graph.

Glazov and Sigmund (2003) also discussed the validity of the nuclear-stopping correction applied in several central experimental papers. It was

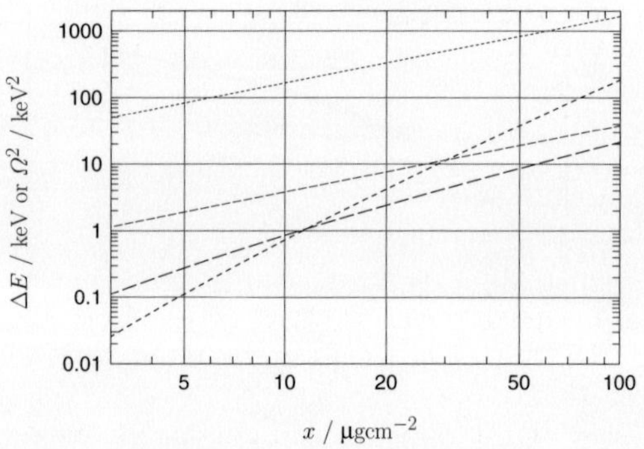

Fig. 15.4. Mean energy loss (thick dashed line) and straggling (thick dotted line) at $\theta = 0$ *versus* target thickness for 320 keV ^{20}Ne ions in carbon for Lenz-Jensen interaction. Thin lines show unrestricted mean energy loss and straggling. From Glazov and Sigmund (2003)

concluded that, where corrections were applied systematically, the adopted procedures were correct from a theoretical point of view, but that the adopted scattering law, based on the LSS Thomas-Fermi interaction, was likely to overestimate the correction and hence to underestimate electronic stopping in the low-speed range where the correction is sizable.

Analysing peak energy losses has the advantage of a comparatively small nuclear-stopping correction. In particular, the high-loss peak for $M_1 > M_2$ is of no concern, and the range of validity of scaling laws is wider than for average energy losses. On the other hand, peak energy losses are more sensitive to the elastic-scattering law and to foil inhomogeneities, and corrections for nuclear scattering are harder to determine theoretically. Moreover, it has been mentioned that measuring peak energy losses imposes high requirements on the monochromaticity of the incident beam.

Measuring average energy losses, while implying bigger nuclear-stopping corrections, is more directly related to the quantity sought. After all, it is the mean electronic energy loss, not the peak loss, that is the quantity of interest in most experiments. Moreover the correction is less sensitive to the scattering law. The high-loss peak should be of little concern if a suitable upper limit is introduced for averaging. However, accurate measurements require reasonably low noise levels.

References

Bohr, N. (1948). "The penetration of atomic particles through matter," Mat. Fys. Medd. Dan. Vid. Selsk. **18 no. 8**, 1–144.

Fastrup, B., Hvelplund, P. and Sautter, C. A. (1966). "Stopping cross section in carbon of 0.1-1.0 MeV atoms with $6 < Z_1 < 20$," Mat. Fys. Medd. Dan. Vid. Selsk. **35 no. 10**, 1–28.

Geissel, H., Lennard, W. N., Andrews, H. R., Ward, D. and Phillips, D. (1984). "Problems of interpreting energy loss data for non-zero emergent angles," Phys. Lett. A **106**, 371–373.

Glazov, L. G. and Sigmund, P. (2003). "Nuclear stopping in transmission measurements," Nucl. Instrum. Methods B **207**, 240–256.

Krist, T., Mertens, P. and Biersack, J. P. (1984). "Nuclear stopping power for particles transmitted through thin foils in the beam direction," Nucl. Instrum. Methods B **2**, 177–181.

Lennard, W. N., Geissel, H., Jackson, D. and Phillips, D. (1986). "Electronic stopping values for low velocity ions ($9 \leq Z_1 \leq 92$) in carbon targets," Nucl. Instrum. Methods B **13**, 127–132.

Ormrod, J. H. and Duckworth, H. E. (1963). "Stopping cross sections in carbon for low-energy atoms with $Z \leq 12$," Can. J. Phys. **41**, 1424–1442.

Skoog, R. (1975). "Elastic interaction of atomic projectiles in an amorphous target," Radiat. Eff. **27**, 53–58.

Williams, E. J. (1939). "Concerning the scattering of fast electrons and of cosmic-ray particles," Proc. Roy. Soc. A **169**, 531–572.

Williams, E. J. (1940). "Multiple scattering of fast electrons and alpha-particles, and "curvature" of cloud tracks due to scattering," Phys. Rev. **58**, 292–306.

16 Range and Range Straggling

16.1 Introductory Remarks

The basic range concepts have been defined in Sect. 2.4. For not-too-low-energy ions, central quantities are the range along the path R and range straggling Ω_R^2 specified in (2.17) and (2.18). The focus of this monograph is on the regime of dominating electronic stopping. It is tempting, therefore, to evaluate the integrals on the basis of stopping cross sections and straggling parameters that ignore nuclear stopping altogether. This approach is justified at high beam energies. The main purpose of this chapter is to specify limitations to this simple approach and to provide suitable corrections to cover the energy regime in which they are comparatively small. Items of concern are

- the influence of nuclear stopping on pathlength,
- the detour factor, i.e., the ratio between projected range and pathlength,
- the validity of the gaussian approximation for the range profile,
- the relative significance of nuclear and electronic energy loss and of angular deflection to range straggling and
- the effect of fluctuating charge states on ranges.

In addition, attention will be paid to

- the extraction of electronic stopping cross sections from range measurements and
- the difference between range and energy-deposition profiles.

Recourse will be made to comprehensive transport equations that also remain valid in the regime of dominating nuclear stopping.

A rough estimate of the importance of nuclear stopping can be found from Table 16.1 which expresses the dimensionless energy variable (10.3) in terms of the mass numbers A_1, A_2 at three characteristic beam velocities. To arrive at these simple expressions one approximates $Z_{1,2} \simeq A_{1,2}/2$ and $Z_1^{2/3} + Z_2^{2/3} \sim (Z_1 + Z_2)^{2/3}$. According to Fig. 10.2 the nuclear stopping cross section has its maximum at $\epsilon \simeq 0.3$. Hence, at 1 keV/u, nuclear stopping is of central importance for $A_1 + A_2 > 100$ and is a significant correction also below this limit. At the Bohr velocity, on the other hand, the corresponding limit is given by $A_1 + A_2 > 1000$, implying that nuclear stopping is just a correction and never dominating. Maximum electronic stopping is found at $\sim v_0 Z_1^{2/3}$. This corresponds to an even higher value of ϵ especially for $Z_1 \gg 1$.

Table 16.1. Values of dimensionless energy variable ϵ governing nuclear stopping at characteristic beam energies.

Beam energy or speed ϵ, (10.3)	
$E/A_1 = 1$ keV	$\dfrac{130}{(A_1 + A_2)^{4/3}}$
$v = v_0$	$\dfrac{3.2 \cdot 10^3}{(A_1 + A_2)^{4/3}}$
$v = v_0 Z_1^{2/3}$	$\dfrac{1.3 \cdot 10^3}{(1 + A_2/A_1)^{4/3}}$

16.2 Transport Equations

The fundamental equation for the distribution $F(E, R)dR$ in total path-length (range along the path) R for an ion with an initial energy E is (Lindhard & al., 1963)

$$-\frac{\partial}{\partial R} F(E, R) = n \int d\sigma(E, w)\Big(F(E, R) - F(E - w, R)\Big). \qquad (16.1)$$

In contrast to (12.18) this is a transport equation of the backward type, where E and R refer to different points in space and time. This type of equation is more suitable for range calculations than are forward equations, but there are limitations, e.g., for layered media for which recourse to forward equations has to be made.

From (16.1) and the normalization $\int_0^\infty dR F(E, R) = 1$ one obtains an integral equation for the average path length $R(E) = \int_0^\infty dR\, RF(E, R)$,

$$n \int d\sigma(E, w)\Big(R(E) - R(E - w)\Big) = 1. \qquad (16.2)$$

From this one can get back to (2.17) under the assumption of small energy loss per collision, i.e., continuous slowing-down, $w \ll E$.

The corresponding equation for the mean projected range $R_p(E)$ reads

$$1 = n \int d\sigma(E, w)\Big(R_p(E) - \cos\phi R_p(E - w)\Big), \qquad (16.3)$$

where ϕ is the scattering angle in the laboratory system in a collision specified by an energy loss w. The relation between $R(E)$ and $R_p(E)$ and other range quantities has been discussed by Lindhard & al. (1963) and Winterbon & al. (1970). In general, relations between range quantities may be derived from the distribution in vector range $F(\boldsymbol{v},\boldsymbol{r})\,\mathrm{d}^3\boldsymbol{r}$ which also obeys a transport equation of the backward type,

$$-\frac{\boldsymbol{v}}{v}\cdot\boldsymbol{\nabla}F(\boldsymbol{v},\boldsymbol{r}) = n\int \mathrm{d}\sigma(\boldsymbol{v},\boldsymbol{v}')\Big(F(\boldsymbol{v},\boldsymbol{r}) - F(\boldsymbol{v}',\boldsymbol{r})\Big). \tag{16.4}$$

Finally, if the initial charge state I of the beam is of significance, the distribution in vector range $F_I(\boldsymbol{v},\boldsymbol{r})$ obeys the transport equation

$$-\frac{\boldsymbol{v}}{v}\cdot\boldsymbol{\nabla}F_I(\boldsymbol{v},\boldsymbol{r}) = n\sum_J\int \mathrm{d}\sigma_{IJ}(\boldsymbol{v},\boldsymbol{v}')\Big(F_I(\boldsymbol{v},\boldsymbol{r}) - F_J(\boldsymbol{v}',\boldsymbol{r})\Big) \tag{16.5}$$

in the notation of sect. 12.3. Equation (16.5) was first given by Burenkov & al. (1992a).

16.3 Simulation Codes

Range profiles can alternatively be determined by computer simulation codes. Existing codes differ from each other in the statistical method and input (Eckstein, 1991). Trajectory simulation codes are conventionally classified into molecular-dynamics, binary-collision and Monte Carlo codes.

Molecular-dynamics codes solve Newton's equation of motion (or its quantal analog). This technique requires considerable computing power and is not much in use for calculating ranges of swift ions.

Binary-collision codes operate on a given – normally non-random – target structure. The prime input is a table of scattering angles and energy losses *versus* energy and impact parameter. The domain of this type of code is the slowing down in a regular crystal lattice, in particular under channeling conditions. Examples are the MARLOWE code (Robinson and Torrens, 1974), ACAT (Yamamura and Misuno, 1985) and CRYSTAL TRIM (Posselt, 1994).

Monte Carlo codes for particle penetration assume a random medium. The prime input is a table of differential cross sections for elastic nuclear scattering and a table of electronic stopping cross sections and, possibly, electronic straggling. In principle the output of such codes should be equivalent with the solutions of transport equations discussed in Sect. 16.2. Minor differences originate in the treatment of soft (distant) interactions. Monte Carlo codes for gaseous targets as well as early codes for solids operate with a distribution of pathlengths between collisions governed by a mean free path. This mean free path is ill-defined for a cross section that becomes singular at zero scattering angle because of the need for truncation. This problem is circumvented in the

TRIM code (Wilson & al., 1977) by operating with a fixed pathlength that is approximately equal to the internuclear distance in the structure. Truncation of the cross section then becomes unnecessary. Nevertheless, cross sections are always truncated. The choice of cutoff angle or energy is less critical for range calculations than for multiple scattering and other phenomena that are sensitive to soft collisions.

16.4 CSDA Range

The total path length in the csda approximation, (2.17), is given by

$$R(E) = \int_0^E \frac{dE'}{N[S_e(E') + S_n(E')]}.$$
(16.6)

Attention needs to be paid to the low-energy portion of the integrand where both S_e and S_n become small. It would seem tempting to approximate $R(E)$ by the 'electronic path length'

$$R_e(E) = \int_0^E \frac{dE'}{N S_e(E')}.$$
(16.7)

One would then have to define a correction for nuclear stopping,

$$\Delta R(E) = R_e(E) - R(E) = \int_0^E dE' \frac{S_n(E')}{S_e(E')[S_e(E') + S_n(E')]}$$
(16.8)

and note that nuclear stopping is unimportant in the Bethe regime where it constitutes less than 0.1 % of the total stopping. This holds approximately down to the maximum of electronic stopping, which according to Table 16.1, lies far above the maximum of nuclear stopping.

Figure 16.1 shows a rough estimate of the correction to the electronic range based upon the approximations

$$S_e \propto E^{1/2} \quad \text{and } S_n = \text{const.}$$
(16.9)

While the correction evidently decreases with increasing ratio S_e/S_n, the approach to zero appears slow. Therefore, a range calculation in the energy regime below the stopping maximum should never ignore nuclear stopping.

A related problem deals with the fact that stopping forces are tabulated only down to $E_0 = 25$ keV/u in ICRU (2005). Figure 16.2 gives an indication of the error made if the lower integration limit in (16.7) is set to this energy. It is seen that the error becomes negligible only above ~ 1 MeV/u. A table of such ranges for 16 ions in liquid water will be given in Chap. 6 of ICRU (2005).

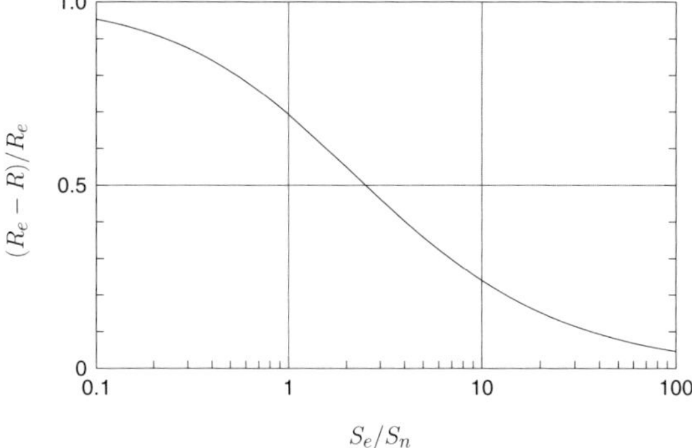

Fig. 16.1. Rough estimate of the effect of nuclear stopping on path length below the electronic-stopping maximum on the basis of (16.9).

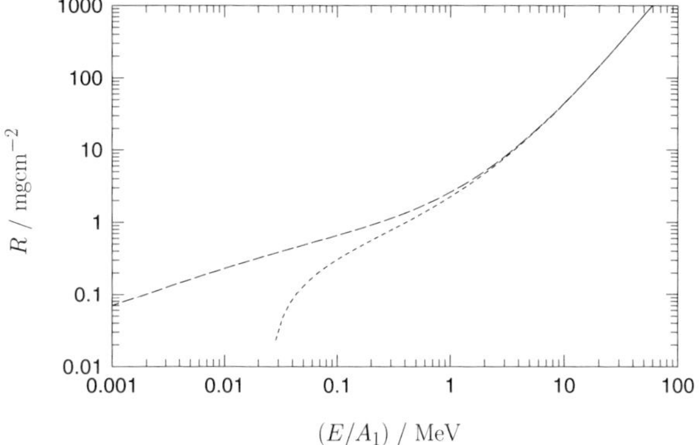

Fig. 16.2. csda range for O - Al. Long-dashed line: Electronic range defined by (16.7), determined from electronic stopping force extrapolated by power law down to zero energy. Short-dashed line: Electronic range determined by integration from 25 keV/u upward.

16.5 Influence of Angular Deflection
on Projected Range

Even though energy loss may be well characterized by the continuous-slowing-down approximation, angular deflection might not be negligible. This has an influence on the projected range and can be estimated on the basis of (16.3), which assuming $w \ll E$ reduces to

$$1 = n\big[S_e(E) + S_1(E)\big]\frac{\mathrm{d}R_p(E)}{\mathrm{d}E} + n\sigma_1(E)R_p(E), \qquad (16.10)$$

where

$$\sigma_1(E) = \int \mathrm{d}\sigma(E,w)(1 - \cos\phi) \qquad (16.11)$$

$$S_1(E) = \int \mathrm{d}\sigma(E,w)w\cos\phi \qquad (16.12)$$

and

$$\cos\phi = \left(1 - \frac{w}{E}\right)^{1/2} + \frac{M_1 - M_2}{2M_1}\frac{w}{E}\left(1 - \frac{w}{E}\right)^{-1/2}. \qquad (16.13)$$

Equation (16.10) was introduced by Schiøtt (1966) to describe ranges of low-energy protons, but the underlying assumptions are well-fulfilled for swift ions in general. It has the general solution

$$R_p(E) = \int_0^E \frac{\mathrm{d}E'}{NS'(E')}\exp\left(-\int_{E'}^E \mathrm{d}E''\frac{\sigma_1(E'')}{S'(E'')}\right), \qquad (16.14)$$

where

$$S'(E) = S_e(E) + S_1(E), \qquad (16.15)$$

This may be written in the form

$$R(E) - R_p(E) = \int_0^E \frac{\mathrm{d}E'}{NS'(E')}\left[1 - \exp\left(-\int_{E'}^E \mathrm{d}E''\frac{\sigma_1(E'')}{S'(E'')}\right)\right]. \qquad (16.16)$$

For not-too-small mass ratios M_1/M_2, one may expand the exponential function in (16.16) and approximate

$$\sigma_1(E) \simeq \frac{M_2}{2M_1}\frac{S_n(E)}{E} \qquad (16.17)$$

$$S_1(E) \simeq S_n(E) - \frac{M_2}{2M_1}\frac{W_n(E)}{E}, \qquad (16.18)$$

where $W_n(E)$ is the nuclear-straggling parameter,

$$W_n(E) = \int_0 \gamma E w \mathrm{d}\sigma_n(w). \qquad (16.19)$$

Then (16.16) reduces to

$$\frac{R(E) - R_p(E)}{R(E)} \simeq \frac{M_2}{2M_1}\int_0^E \frac{\mathrm{d}E'}{E'}\frac{S_n(E')}{S'(E')}\frac{R(E')}{R(E)}, \qquad (16.20)$$

which determines the detour factor $R_p(E)/R(E)$.

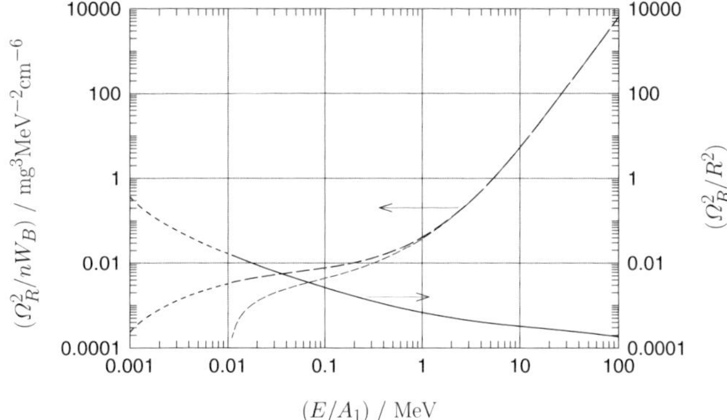

Fig. 16.3. Range straggling evaluated from (2.18) taking into account only electronic stopping and straggling. Thin dashed line: Integration from 0.01 MeV upward. Thick lines based on integration of curves extrapolated to lower energy.

16.6 Range Straggling

Figure 16.3 shows results for csda range straggling corresponding to Fig. 16.2 for the mean range. However, widths of range profiles can be quite sensitive to nuclear straggling and angular scattering at comparatively low energies.

The effect of fluctuating charge deserves attention in high-energy ion ranges. Most important are effects on range straggling. If energy-loss straggling is taken into account, the variance of the range profile may be estimated on the basis of (2.18).

Burenkov & al. (1992c) solved (16.5) numerically for B in Si and found straggling to be increased by 30-40 % over an energy range from $\sim$ 10 to 100 MeV/ion. Burenkov & al. (1992b) performed Monte Carlo simulations and found very pronounced differences from results of the TRIM code that neglects the effect of charge fluctuation.

Transient effects on the mean range in case of pronounced deviations of the initial charge state from equilibrium are possible. However, for heavy ions, fairly thin foils are needed to detect such effects in stopping experiments. Equivalent effects in range must be even harder to identify.

16.7 Extraction of Electronic Stopping Cross Sections from Range Measurements

Stopping data for heavy ions have frequently been determined on the basis of ion ranges. If an $R(E)$ dependence is measured over an adequate energy interval, the stopping force can be found by differentiation. In the simplest form, such a procedure does not take into account nuclear scattering and stopping. Equation (16.10) offers an operational procedure to explicitly take into account this effect in the energy range where it is small. Indeed, rewriting it in the form

$$n\big[S_e(E) + S_1(E)\big] = \frac{1 - n\sigma_1(E)R_p(E)}{\mathrm{d}R_p(E)/\mathrm{d}E}, \tag{16.21}$$

one finds the standard form

$$nS_e(E) = \frac{1}{\mathrm{d}R_p(E)/\mathrm{d}E} \tag{16.22}$$

in the limit of negligible nuclear scattering, but (16.21) offers explicit corrections for both angular deflection and nuclear energy loss. As in section 15 the resulting electronic stopping force depends on the adopted input describing nuclear collisions.

16.8 Ranges of Low-Energy Ions

In the regime of dominating nuclear stopping, theoretical treatments need to make recourse to transport equations. In accordance with the discussion in Sect. 10.2, it is common – and most frequently justified – to decouple electronic from nuclear collisions. Ignoring angular deflection in electronic collisions, this leads to (Lindhard & al., 1963)

$$1 = nS_e(E)\frac{\mathrm{d}R_p(E)}{\mathrm{d}E} + n\int \mathrm{d}\sigma_n(E,w)\Big(R_p(E) - \cos\phi R_p(E-w)\Big) \tag{16.23}$$

and similar relations for other range quantities.

Numerous alternative statistical formulations of the range problem may be found in the literature, and a variety of tools is available for their solution. Most of them have been developed with applications in ion implantation in mind, i.e., within the energy regime in which the simple estimates mentioned above are not necessarily valid. Efficient methods are available to determine averages, especially of first and second order, on the basis of analytical estimates and numerical evaluation of some integrals. Reliable estimates of range profiles are commonly performed either by numerical solution of the transport equation or by Monte Carlo simulation with equivalent input.

References

Burenkov, A. F., Komarov, F. F. and Fedotov, S. A. (1992a). "The ion charge fluctuation effect on impurity depth distributions for high-energy ion implantation — the backward transport equation-based simulation," phys. stat. sol. B **169**, 33–49.

Burenkov, A. F., Komarov, F. F. and Fedotov, S. A. (1992b). "The ion charge state fluctuation effect during high energy ion implantation: Monte-Carlo simulation," Nucl. Instrum. Methods B **67**, 35.

Burenkov, A. F., Komarov, F. F. and Fedotov, S. A. (1992c). "The transport equation approach for the simulation of charge state fluctuation effects during ion implantation into solids," Nucl. Instrum. Methods B **67**, 30–34.

Eckstein, W. (1991). *Computer simulation of ion-solid interactions* (Springer-Verlag, Berlin).

ICRU (2005). "Stopping of Heavy Ions," J. ICRU to appear.

Lindhard, J., Scharff, M. and Schiøtt, H. E. (1963). "Range concepts and heavy ion ranges," Mat. Fys. Medd. Dan. Vid. Selsk. **33 no. 14**, 1.

Posselt, M. (1994). "Crystal-TRIM and its application to investigations on channeling effects during ion implantation," Radiat. Eff. Def.Solids **130**, 87–119.

Robinson, M. T. and Torrens, I. M. (1974). "Computer simulation of atomic-displacement cascades in solids in the binary-collision approximation," Phys. Rev. 5008–5024.

Schiøtt, H. E. (1966). "Range-energy relations for low-energy ions," Mat. Fys. Medd. Dan. Vid. Selsk. **35 no. 9**, 1–20.

Wilson, W. D., Haggmark, L. G. and Biersack, J. P. (1977). "Calculations of nuclear stopping, ranges and straggling in the low-energy region," Phys. Rev. B **15**, 2458–2468.

Winterbon, K. B., Sigmund, P. and Sanders, J. B. (1970). "Spatial distribution of energy deposited by atomic particles in elastic collisions," Mat. Fys. Medd. Dan. Vid. Selsk. **37 no. 14**, 1–73.

Yamamura, Y. and Misuno, Y. (1985). "ACAT," Technical report IPPJ-AM-40, Institute of Plasma Physics, Nagoya University.

17 Concluding Remarks

With reference to Fig. 4.1 one may first note that significant progress has been made over the past few years in the understanding of heavy-ion stopping over the entire velocity range covered in the graph. A fairly comprehensive survey has been given in a recent conference (Andersen and Sigmund, 2002). The highly relativistic regime is well-covered by the theory of Lindhard and Sørensen (1996). Further down in speed lies the domain of the binary theory and of the CasP code. While the range of validity of the binary theory reaches well below the stopping maximum, that of the CasP code is more restricted because of the lack of shell and Barkas-Andersen corrections. The theory of Arista (2002) has been geared toward low projectile speeds but is becoming successful also around and above the stopping maximum. The theory of Maynard & al. (2001) is particularly successful in connection with hot (plasma) targets.

Key elements in this area are proper account of the Bloch theory and an understanding of the Barkas-Andersen effect as well as projectile screening. At the same time, several features familiar from the Bethe theory are less prominent for heavy ions. This is particularly true for deviations from Bragg additivity, which become less pronounced as one leaves the Bethe regime.

Some clarification has been achieved regarding oscillatory effects, i.e., Z_1 and Z_2 structure, in particular with regard to behavior as a function of speed, but more attention is needed to the quantitative behavior. Arista's theory comes close to Z_1 oscillations, but it applies to free-electron systems only and hence does not make statements about gases and insulators. Binary theory elucidates Z_2 structure, but that structure is most pronounced in the low-speed range for which predictions of the binary theory become uncertain.

Significant progress has also been achieved in understanding and quantifying energy-loss straggling with and without charge exchange, although much more work is needed. Less progress has been reported in more traditional areas such as the analytical theory nuclear scattering, multiple scattering and range profiles. Efficient tools and computer codes handling these aspects have been available for many years.

References

Andersen, H. H. and Sigmund, P., eds. (2002). *Stopping of heavy ions – STOP 01* (Nucl. Instrum. Methods B 195).

Arista, N. R. (2002). "Energy loss of heavy ions in solids: non-linear calculations for slow and swift ions," Nucl. Instrum. Methods B **195**, 91–105.

Lindhard, J. and Sørensen, A. H. (1996). "On the relativistic theory of stopping of heavy ions," Phys. Rev. A **53**, 2443–2456.

Maynard, G., Zwicknagel, G., Deutsch, C. and Katsonis, K. (2001). "Diffusion-transport cross section and stopping power of swift heavy ions - art. no. 052903," Phys. Rev. A **63**, 052903-1–14.

Subject Index

Springer Tracts in Modern Physics

Druck: betz-druck GmbH, D-64291 Darmstadt
Verarbeitung: Buchbinderei Schäffer, D-67269 Grünstadt